HOME DAIRYING

*This book is dedicated
to the memory of my late mother,
Mrs Sydna Ann Smith.*

HOME DAIRYING

Katie Thear

Cheesemakers' Journal   Ashfield MA

PREFACE TO THE AMERICAN EDITION

We are extremely pleased to be able to offer for publication Katie Thear's book, Home Dairying. This is an excellent work which is filled with detailed information on the selection and keeping of family cows, goats, and dairy sheep. Mrs. Thear describes the conditions necessary for producing your own cream, butter, yogurt, and cheese, whether you have a small home dairy or only a kitchen at your disposal.

We know that this book will be a welcome addition to any home or small scale farm cheesemaker. For any readers who may have any questions relating to home dairying please feel free to address them to us, here in Ashfield, Massachusetts.

Bob & Ricki Carroll
Publishers
Cheesemakers' Journal
Ashfield, MA 01330

CONTENTS

Preface 7

1 Dairy Animals 9

Dairy cattle 9
Dairy goats 10
Dairy ewes 11
Milk production 12
Milking 13

2 The Dairy 17

The traditional dairy 17
Adapting an outbuilding as a small dairy 17
The kitchen as a dairy 19
The conditions for dairying 19

3 Milk 21

Types of milk 21
Milk treatment in the dairy 23
Packaging and selling milk 24
Recipes using milk 24

4 Yoghurt 26

Making yoghurt at home 26
Producing flavoured yoghurts 27
Some problems 28
Small-scale commercial yoghurt production 29
Others forms of yoghurt 30
Storing yoghurt 31
Cooking with yoghurt 31
Recipes using yoghurt 32

5 Cream   33

Cream separation   33
Storage of cream   35
Pasteurizing cream   35
Clotted cream   35
Making clotted cream at home   36
Sour cream   36
Ice cream   36
Selling cream   39

6 Butter   40

Buttermaking   40
Storing butter   45
Buttermaking problems   45
Butter recipes   46

7 Cheese   47

Types of cheese   47
The elements of cheesemaking   48
Cheesemaking equipment   53
The principles of hard cheesemaking   55
Growing moulds   60
Colouring cheese   62
Keeping records   62
Pests and problems   63
Selling cheeses   65

8 Cheesemaking Recipes   66

Recipes using cheese   87
Cheesecakes   88

Useful addresses   90
Bibliography   93
Index   95

PREFACE

When I wrote my first book on home dairying in 1978 I little thought that there would prove to be such an interest in the subject. The period between the early 1950s and late 1960s saw the demise of many small farms and with them went many of the traditional practices of cheese and buttermaking on a small, local scale as agriculture became large, intensive and specialized. The factory took over, not only the farm, but many of its products. The creamery replaced the farmer's wife, manufacturers stopped producing equipment for small-scale production, and many of the tools previously in everyday use ended up as 'dairying bygones' in museums or antique shops.

I grew up with dairy animals, and milk churning and buttermaking were regular occurrences in my old home on the Lleyn peninsula in north-west Wales. Here progress came much later than in many other areas. There was no electricity or piped water until the late 1950s and most of the small farms had their own cows and produced their own dairy products. Now, none of them do and much of the area is given over to the tourist trade. The only thing that has remained unchanged is that the people still speak Welsh. On a recent visit, I went to see the farm, Plasymhenllech, where my mother learnt cheesemaking at the age of eighteen in 1917. Nothing remains to show that there was even a dairy there. Yet, once it was a centre of traditional dairying practice that went back many generations.

In recent years it has been the new small farmers, often people who have moved from urban areas, who have revived dairying crafts. When I began to produce a small farming magazine in 1975, I realized that while there were many people interested in finding out more about small-scale dairying much of the traditional knowledge was disappearing before it could be recorded, so I spent three years in research, visiting small farms all over the country, collecting cheese recipes, and going on dairying courses which provided up-to-date knowledge and techniques. As a result I published *The Home Dairying Book* (Broad Leys Publishing Co., 1978) which was cheaply produced and not distributed through the bookshops, yet proved amazingly popular. Fairly soon other books began to appear on the same subject and I experienced a mixture of irritation and flattery when I saw how frequently my researches, writings and recipes were being plagarized.

Now the original book has been completely rewritten and augmented with a great deal of new material.

My thanks are due to all the farmers' wives, goatkeepers and many others who have helped to make this book possible by contributing their recipes and advice.

Katie Thear,
Widdington, 1983.

DAIRY ANIMALS

The most common dairy animals are cows, goats and sheep, although ass's milk has been much utilized in the past. Cleopatra is reputed to have bathed in it, and as recently as the turn of the century it was fetching a good price in London street markets. The cow stands supreme in the dairy stakes, although recent years have seen a remarkable increase in the number of people keeping goats. This is a cheering trend; for far too long goats have had a quite unjustified prejudice against them. Milking sheep are still a minority, but here again there is a surprising interest, and it should perhaps not be forgotten that the famous Roquefort cheese of France is made from sheep's milk. In the last few years there has been considerable emphasis on cross-breeding of European Friesland sheep with British breeds in order to produce hardier crosses which are also suitable for meat lamb production.

Dairy cattle

Dairy cattle are to be found in areas where the pasture is of a good quality. In Britain this is generally in the north and western areas where the relatively high rainfall produces lush growth. The drier south-eastern counties are where most of the cereal crops are grown. This pattern is repeated in most areas of the world, with the dairy herds being concentrated where the best grazing is available. The Normandy area of France has long been associated with dairying, and before the advent of the European Common Market, New Zealand, with its mild climate, adequate rainfall and good pastureland, produced a high proportion of cheap butter for exporting to Britain.

The main dairy breeds are:

Jersey The Jersey cow's milk is high in carotene and butterfat, giving it its characteristic golden yellow colour. It is a small cow, docile and easy to handle, making it an ideal house cow. It is very common in the USA.

Guernsey The Guernsey too has milk rich in butterfat, and of a golden yellow colour. It is bigger than the Jersey although just as docile.

Ayrshire The Ayrshire was originally bred in Scotland. It is widely kept as a dairy animal, giving rich milk.

Friesian This is the most widely kept dairy cow in commercial herds. The volume of milk is very high, but with a lower butterfat content than that of the Channel Islands breeds or the Ayrshire.

Dairy Shorthorn This is a good dairy cow which is also useful as a producer of beef.

Normandy The Normandy is a large cow, particularly common in France, where it is the premier dairy animal.

Swiss Brown This is a native of Switzerland and although not so common in Britain, has been exported and bred in large numbers in the USA.

Holstein This was bred from the Friesian and produces a hugh volume of milk with a low butterfat content. It is a large cow

Fig. 1 *Some breeds of dairy cow*

suitable for the large-scale milk producer, and very common in the USA.

South Devon   This is a large cow, popular in the west country in Britain. It produces milk with a high butterfat content and is also useful for beef.

Welsh Black   The Welsh Black was once the premier dairy cow in Wales and was the breed kept by my parents and their compatriots. In recent years it has been developed more as a beef animal.

Dexter   The smallest British cow and a hardy animal that needs far less pasture than the average cow. It is good for beef as well as milk, but can be bad-tempered. There may also be problems associated with breeding and availability is fairly limited.

There are, of course, many other breeds of dairy cattle, with each area of the world having its own indigenous or early introduced breeds. The Kerry, for example, is claimed to have existed in Ireland for thousands of years while the Swedish Red is popular in Sweden, yet not often seen outside that country. The most widely kept breeds on a commercial scale are the Friesian, the Holstein and the Jersey. These are readily available in many countries.

Dairy goats

On a world scale, goats are far more numerous as livestock than cows, but in most African and many eastern countries they are kept as meat producers rather than as dairy animals. Commercial dairy goats are more common in Switzerland, France and the Mediterranean areas of Europe than they are in Britain. It is only in recent years that there has been an increase in the number of goats kept for commercial purposes. This has no doubt come about because of the growing interest in goats' milk products and many health food shops and delicatessens now sell both goats' milk and yoghurt. Most of these come from small enterprises and a high proportion of goat-keepers in this country still milk by hand, although milking machines are available for goats.

The main breeds are:

Saanen   Originally from Switzerland, the Saanen is generally quiet and docile, with a good record of milk yield.

British Saanen   The British Saanen is in some respects an improved Saanen, and is both bigger and with more prolific milk yields.

Toggenburg   The Toggenburg originated in Switzerland, and is a small breed which is

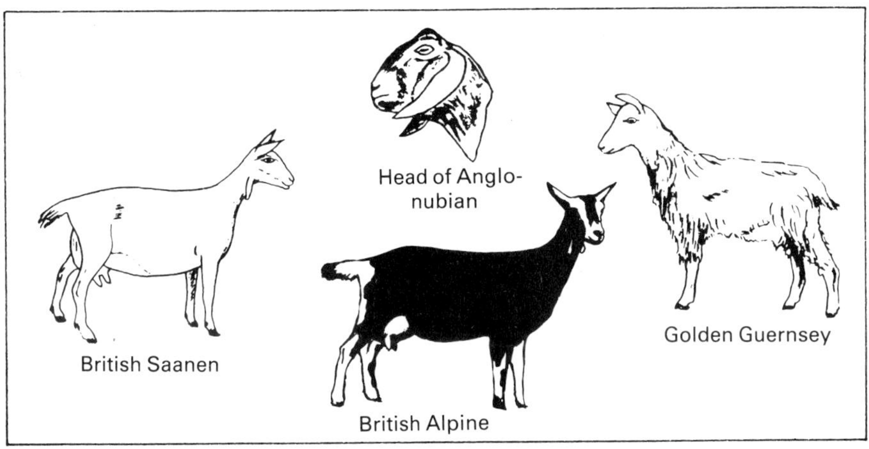

Fig. 2 *Some breeds of dairy goat*

generally docile. Its milk yield is not as high in this country as in its country of origin, partly due to the past shortage of imported blood stock.

British Toggenburg This is larger and darker than the Toggenburg, with a good record of milk yield.

British Alpine This breed was developed in Britain, originally from the same stock as the British Toggenburg, but incorporating the black colour from a different line. They are good milkers.

Anglo-Nubian This is a breed that is distinctive, not only for its Roman nose, but for the highest butterfat content of any goat's milk. It was bred in Britain by crossing native and cross-bred goats with males imported from the east.

British This can mean any cross-bred goat of varying type, including scrub goats. There are some very good milkers to be found among them, but when buying a goat described as British, it is important to buy stock from 'milk-recorded' parents with a history of good production levels.

Golden Guernsey This, as the name implies, originated from stock on the island of Guernsey and is golden in colour. It is becoming increasingly popular, although still limited in numbers.

English Guernsey Since 1975 the British Goat Society has made allowances for selected outcrossings between Golden Guernsey females with Saanen or British Saanen sires, and stock can gain entry after being mated back to Golden Guernsey or English Guernsey sires for three generations.

The United States has Saanens, Toggenburgs, Alpines and Nubians which have originated from European stock and have subsequently been developed and crossbred. In addition, they have the La Mancha, a short-eared South American goat which has been increasingly bred as a dairy goat in recent years. Australia has most of the European breeds which were originally imported by the early settlers. There is also a large population of feral or semi-wild goats which are the offspring of early introductions, and which are regularly culled for the meat trade.

Dairy ewes

What is a milk sheep? Any sheep that has just lambed will produce milk, but there are obviously characteristics that make certain sheep more suitable as dairy animals than others. Long legs for access to the udder is an obvious advantage, and the traditional Mediterranean breeds have this feature. In fact, some of them look more like goats than sheep. Milk yield is another crucial factor and perhaps the highest level is held by the Fries Melkschaap or Frieslands which are found in Holland and North Germany, and which have been imported into Britain. They yield an average of 140 gallons in a 250-day lactation, and the milk has a butterfat content of 6–7%.

In Britain, in recent years, there has been a certain amount of cross-breeding of Frieslands with local breeds with the aim of up-grading the local breeds for dairy production, without losing their basic hardiness. There is now a British Friesland Society which looks after the interests of breeders and which will provide information on sources of stock. The address is in the reference section. Breeds of milk sheep found in different parts of the world are:

East Friesland	Germany
Dutch Friesland	Holland
British Friesland	Britain
Lacuna	
Larzac	France
Pramenka	Yugoslavia
Daglic	
Sakiz	Turkey
Awassi	Israel
Chios	Greece
Sopravissina	Italy
Barbary	Mediterranean area

Dairy ewes are most uncommon in the USA but there is a growing interest in them among dairy specialists.

Milk production

It is beyond the scope of this book to go into the specific details of feeding, housing and general management of dairy animals. There are many good books which cover the practical care of farm livestock and these are listed in the reference section at the back of the book.

As far as milk production is concerned, all dairy animals need enough food to cover their basic metabolic needs, as well as an extra ration which is related to the amount of milk they are producing. The former is referred to as a 'maintenance ration' while the latter is a 'production' ration. In commercial enterprises this is worked out exactly so that the appropriate amount of feed is given, while maintaining a cost-effective operation. In summer, grass provides an important proportion of the daily intake. In winter, when the grass has stopped growing, preserved grass in the form of hay (dried grass) or silage (pickled grass) is made available. A dairy concentrate ration made up of various grains is also given all year round to cater for the demands of milk production. Most dairy farmers also make hay available right through the year, but obviously give less in summer than in winter. Fodder crops such as kale are frequently grown for winter feeding. Mineral licks are provided to ensure that essential minerals and trace elements are available to keep the stock healthy. A constant supply of water is essential for all dairy animals. While in full lactation, an animal can drink several gallons a day.

Regular care to prevent health problems is a priority and all animals on pasture are regularly wormed with vermifuges to kill internal parasites. In addition, external parasites such as lice, mites, ticks and warble flies are kept at bay by the use of veterinary powders and washes. Immunizations to prevent diseases are administered and regular tasks such as keeping the animals' feet in good condition are undertaken. This involves trimming back the hooves to prevent the nails curling over. In the case of dairy ewes the regular running of a flock through a footbath is also carried out to prevent footrot, a condition which affects sheep and, to a lesser extent, goats. Sheep are also regularly dipped or immersed in insecticidal solutions to prevent scab, a particularly nasty parasite, as well as a range of other external parasites.

The prevention of mastitis is one of the most important aspects of dairy farming,

Fig. 3 *Fore-milk cup with detachable examination dish*

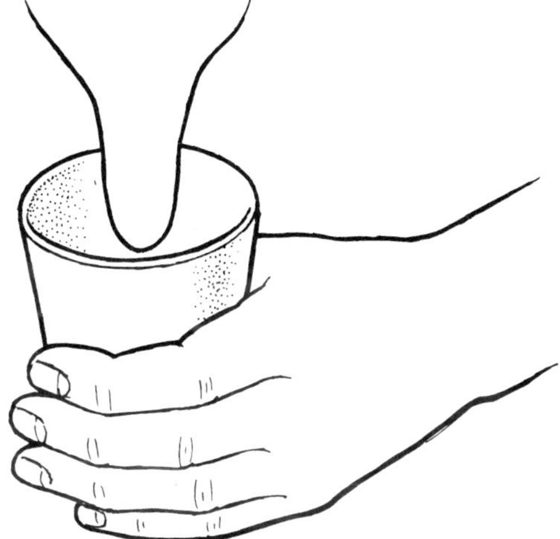

Fig. 4 *Teat dipping*

particularly in heavy-yielding animals where a strain is placed on the udder. Mastitis is a nasty condition of the udder that must be continually watched out for by checking the fore-milk in a strip cup (Fig. 3). Its presence is indicated by spots of blood and white or yellow spots. It is caused by either staphylococcus or streptococcus bacteria which gain entry into the udder through injury, or via the teat canal itself. Stringent standards of cleanliness in milking practice are necessary to avoid the possibility of mastitis. Any cut on the udder or teat must be treated immediately with an antiseptic lotion or cream. There is an excellent case for banning that pernicious barbed wire that has been responsible for so many accidents. The practice of teat dipping (Fig. 4), where each teat is dipped into an anti-bacterial solution, is one way of reducing the risk of infection. All milking

machines and equipment should be cleaned to exacting standards to provide hygienic conditions and operatives should follow a code of practice to ensure personal cleanliness.

If left untreated, mastitis can develop into a chronic infection which lowers the general health of the animal to such an extent that she will be prone to attacks from other diseases. It may also flare up into a severe fever which will permanently damage her constitution. Even if she recovers, her udder may have been permanently damaged and partly blocked by scar tissue. Mastitis is also one of those conditions which has a tendency to recur. If it is detected, antibiotics are prescribed. These are administered into the teat canal by means of an inter-mammary syringe (Fig. 5). Milking takes place as usual, but the milk is discarded until the condition has cleared up. Once the course of penicillin has been completed, three clear days are allowed to elapse before the milk is used for consumption. There are regular tests carried out on cow's milk to detect the presence of antibiotic residues or other contaminants and legal penalties may be incurred by farmers who do not comply with the regulations.

Milking

A cow's udder is made up of four quarters, while that of the goat and sheep has two sections. Each quarter (Fig. 6) is separate and is held in place by suspensory ligaments. The udder is composed of secretory tissue made up of a collection of alveoli to which the milk constituents are delivered by the bloodstream. From the alveoli, the milk drains via the ducts, into the milk cistern above the teat. A fold of tissue separates the udder cistern and teat canal, and the end of the teat has strong muscles at the opening. When the teat is squeezed at the top, the circular fold of tissue closes, shutting off the cistern from the canal and with pressure exerted downwards the milk

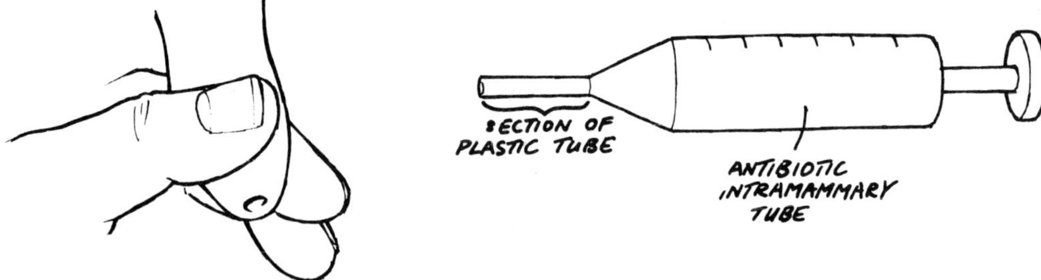

Fig. 5 *Administering an antibiotic after milking out*

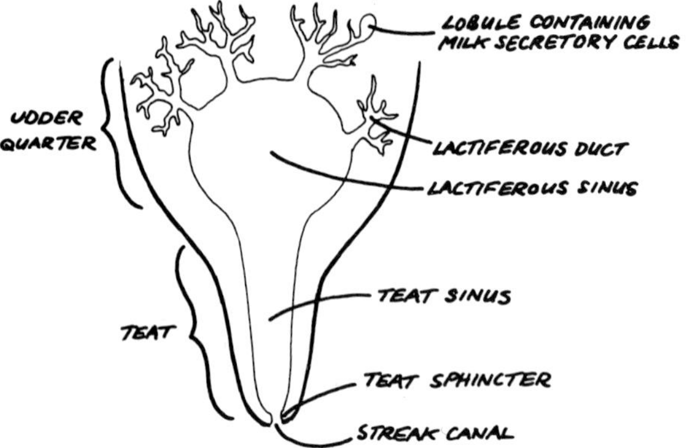

Fig. 6 *Diagram of an udder quarter*

spurts out through the teat opening.

If the interval between milkings is too long, pressure builds up in the udder and the milk constituents are reabsorbed into the blood stream, with a consequent decline in overall milk yield. For high-yielding cows in full lactation the optimum time limit has been demonstrated to be nine hours, but as this would entail milking three times a day it is not a common practice. The normal practice is for milking to be carried out in the morning and the evening, at the same times each day so that the animals become used to the same routine.

A dairy animal can, to a limited extent, control the 'let-down' of milk. What the experienced milker does is to encourage her. Sometimes just the sound of the milk pail rattling is enough to stimulate the milk flow, and the gentle washing of the udder can have the same effect. It used to be a common practice to sing to the cow, and there are still songs and rhymes around that testify to this gentle and worthy tradition.

Little cow, little cow,
Let down thy milk,
And I will give thee
a fine gown of silk.

It is in fact a reaction to the hormone oxytocin from the pituitary gland that allows the milk to descend from the cavities of the alveoli into the milk ducts leading to the gland cistern and teat canals. This process is also inhibited by a hormone, and the relationship and degree of trust between the animal and her owner is a most important aspect. Most dairy animals are milked by machine in specially constructed milking parlours built to comply with the hygiene requirements of the dairying regulations. A substantial minority of people milk by hand if they have just a few animals to provide milk for the home, but it must be added that there are also many goatkeepers who handmilk despite the fact that they are producing commercial dairy products. The main reason for this is that the dairying regulations in relation to goats are far less stringent than they are with cows where milk can only be marketed through the auspices of the Milk Marketing Board. It is also undeniable that the official bodies have tended not to take goats seriously in Britain and have therefore overlooked them. This situation may not continue much longer, particularly as the numbers of people buying dairy goat products increases.

Handmilking is not difficult but, like so

many things, it requires practice. The ideal way of learning to milk is to get an experienced person to demonstrate, and then to practise, under his or her watchful eye, on a well-tempered patient cow or goat which is nearing the end of its lactation period. It is not always possible to have the ideal, but it is at least possible to learn the theory before going out and buying a dairy animal. There are several organizations running dairying courses where practical tuition of this nature is available.

Before milking, wipe the udder carefully with a clean udder cloth dipped in warm water with a little sterilizing agent added. Pat dry to prevent moisture running down the teat into the bucket. Wash hands thoroughly and keep nails cut short and filed smooth.

Ensure that the milking pail is not used for any other purpose, but is kept purely for milking. After use rinse in cold water first then wash thoroughly in boiling water. Stainless steel is the best material as it does not become scratched and so harbour bacteria in the way that plastic may. Galvanized metal is suitable provided there are no scratches but it does not last as long as stainless steel.

Squirt the fore-milk from each teat into a strip cup – or any clean container that will show up abnormalities in the milk. Examine carefully for specks of blood or anything unusual that could indicate mastitis or some other condition of the udder. This examination is really vital and is required by law for those who sell their milk. The fore-milk is rich in bacteria anyway and should always be discarded before milking proper. The traditional practice was to give this milk to the farm cats who were usually ready and waiting at milking time.

Milk as quickly and as efficiently as you can, so that as little dust, hair or any other foreign matter gets into the milk. Filter the milk immediately by using a filter unit equipped with disposable filter discs. This removes any matter or hair which may have fallen in during milking.

You can milk one teat at a time, but it is much quicker to milk two, and with practice this is not difficult. With a cow, it does not really matter which combination of two of the four teats you milk at a time. Hold the tests gently but firmly, with the thumbs just below the bag. Inside the teat is a canal leading from the milk cavity or gland cistern to the outside. This canal fills with milk and in order for it to come out, the top of the canal needs to be closed off by applying gentle pressure between the thumb and index finger. Close the rest of the hand around the teat, still keeping the top of the canal constricted, and the milk is forced outwards. Release pressure between thumb and forefinger, allowing milk to fill the canal again, and continue as before. Alternate this action between the two hands. Take great care not to hold the teat too high in case the tissue in the bag is damaged by being constricted between the finger and thumb. It is not necessary to move the arms when milking, nor to pull the teats and these constitute the most common mistakes when first learning.

Stripping is the process of getting the last milk out of the teat by squeezing the top, then gently sliding the thumb and index finger down the length of the teat. The udder is a delicate mechanism and the practice of stripping should be kept to a minimum.

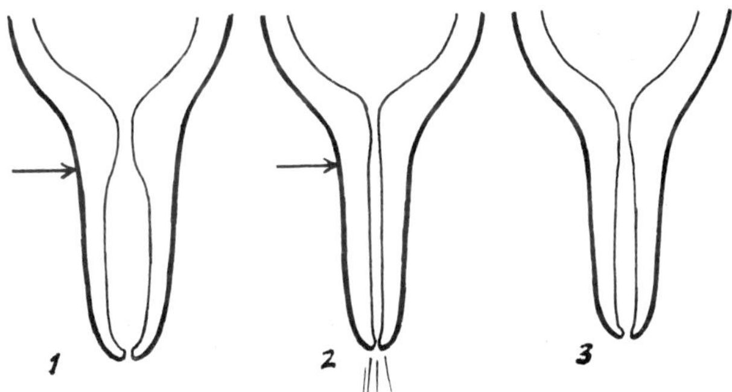

Fig. 7 *Milking sequence*
1 *Exert pressure with index finger and thumb.*
2 *Maintain pressure here, squeeze teat with remaining fingers and hand.* **3** *Release pressure and allow teat canal to fill up again.*

Milk cans with lids are available in a variety of sizes. Unfortunately the traditional, galvanized milk churn is no longer being manufactured, since bulk tanker transport of milk has become predominant.

Traditionally sheep were milked from behind but this was a risky business as the bucket was right in the path of any droppings that might appear without warning. In France, and other Mediterranean areas where milk sheep have been kept for many generations, the practice was to milk by standing astride the sheep, facing its tail. Some people milk sheep from the side and either kneel on the ground or sit on a low stool with the sheep standing on a platform.

The teats of the ewe are shorter than those of a cow and goat, and practice is needed to grasp the teats firmly without pulling. Men, with their larger hands, will find it even more difficult. The base of the teat where it meets the bag, has a circle of areolae which feel like raised spots, and it is here that pressure needs to be applied. Most dairy ewes in commercial flocks are machine milked.

The principle of machine milking is that the vacuum produced in the teat cups imitates the sucking action of the young and milk is drawn out in response to this. It is important that the vacuum is not too high otherwise there is a risk of damaging the delicate lining of the teat.

For cows, the level should be 38 cm (15 in) mercury, which is equivalent to 3.3 kg (7½ lbs) pressure per square inch, and regular checking should be carried out to ensure that this is not exceeded. If it is exceeded, the danger is that the teat cups creep up and pinch the bottom of the udder, and possibly damage the tissues. For goats and sheep the pressure will need to be lower.

The pipeline system of machine milking where milk from each animal is piped direct to a central collection point, is not appropriate for the single cow owner, or for those with a couple of goats. For commercial goat herds however, it may well have a place, and in recent years, pipeline units have become available for goats and sheep as well as cows. The bucket system, where milk from one or two animals goes into a specialized bucket, is more appropriate for the smallholder, but it is still very expensive unless one can find a second hand unit.

It is customary to keep a record of milk yields to keep track of the progress of individual animals and their nutritional needs as well as to give a guide to the best producers and potential. When calves, kids or lambs are sold, the prospective buyer may wish to know what the milk yield of the mother has been. Milk is weighed in pounds or kilogrammes and it is useful to have scales hanging up in the milking parlour for this purpose. Many people also have a calendar hanging up, so that the yields are recorded on the appropriate dates, but a more reliable and permanent record is to write them in a book kept specifically for this purpose. The British Goat Society has milk-recording cards or books for sale.

2

THE DAIRY

The traditional dairy

It was not just farms which had dairies in the past. Most country houses and many cottages had them, for dairying was an important part of everyday life. There was no daily pint of milk delivered on the door-step and people had to provide for themselves. Even houses on the edge of towns often had their own small butteries where milk would be left to settle so that the cream could be skimmed off for butter. A large farm, with a herd of cows would have had the dairy in a building separate from the farmhouse. A smaller household with perhaps just one or two cows for family use, usually had the dairy incorporated on the north-facing side of the house, where it was cool.

Figure 8 shows nineteenth-century examples of a small household dairy and a farm one. Figure 8 (a) is the plan of a small house with a kitchen, parlour and dairy occupying the ground floor which has three bedrooms above. The pantry, scullery and fuel store are in a single-storey lean-to. Water is from a well and there is an outside privy and cesspool. It is interesting to note the relative sizes of the rooms, with the kitchen bigger than the parlour, importance being attached more to the utility aspect rather than leisure. Figure 8 (b) is the plan of a dairy on a 100-acre farm of the same period. It is a completely detached building facing north. The milk was brought in churns along the corridor to the setting room which is equipped with a creaming apparatus consisting of cold water fittings and jackets. This was obviously a fairly advanced dairy for its time because most from that period would have had shallow setting pans on shelves to hold the milk while the cream rose and settled on the surface.

The setting room also doubled as a butter store, with the butter placed on slate shelves for coolness. The main dairy area is equipped with a washing trough, a churn and a butter worker for removing surplus water from the butter. There is also a table and an early refrigerator. The floor is stone-flagged, with a channel for draining away water and the boiler house provides heating facilities for the provision of hot as well as cold water. The verandah shown in the illustration would have been for draining the cans and other utensils on the slatted bench in the fresh air.

Since that period dairying has now largely become removed from the farmer who merely produces the milk. The milk is collected by bulk tanker and taken to the bottling plant.

Adapting an outbuilding as a small dairy

In recent years, there has been a resurgence of interest in home dairying, and people have once more begun to look around for a suitable place in which to carry out the various activities. The problem is that most of the traditional dairies no longer exist.

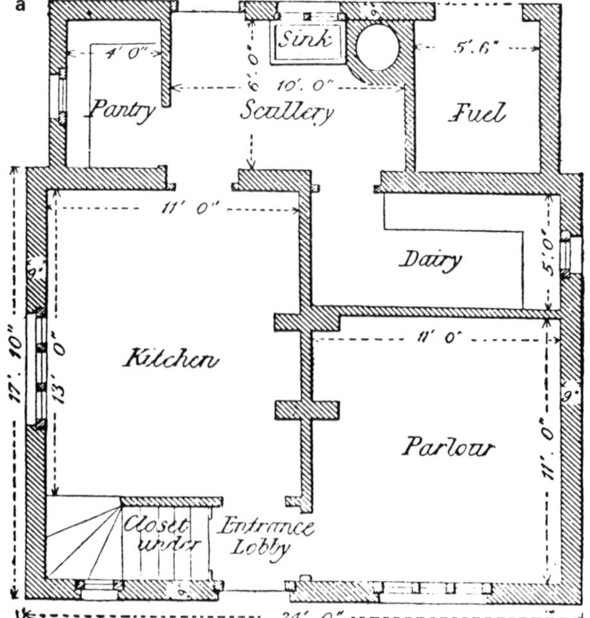

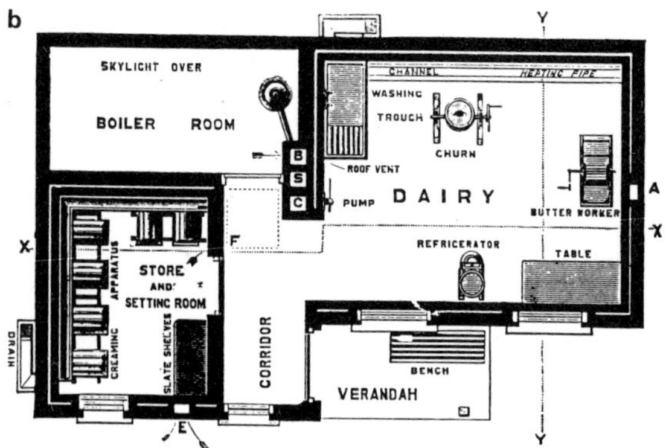

Fig. 8 *(a) and (b). Illustrations from* Farm Buildings. Their Construction and Arrangement *by A. Dudley Clarke (B.T. Batsford, 1891): (a) shows an example of a small household dairy incorporated into the dwelling; (b) shows an example of a farm dairy housed in a separate building.*

Many of those that were in older houses have been turned into bathrooms or extended as extra living accommodation. In my present home the main living room is where the old buttery was. Many of the farm dairies have also disappeared, although there are still a few preserved as museum pieces. The one at Easton Farm Park in East Anglia is a perfectly preserved Victorian dairy and is well worth a visit.

Ideally, the handling and manufacture of dairy products should be in a purpose-built dairy, but the cost of such a building is beyond the budget of most people whose activities are on a small scale, and it is often the kitchen that has to act as a make-do dairy.

If you are fortunate enough to have an existing building or room that is ripe for conversion, the important features in or associated with a dairy are:

Cool, airy aspect   There should be windows to provide light, but it is better to have an air extractor fitted in the glass, rather than have the windows open, to discourage an influx of flies. Space heating in any form is not required and the introduction of an electric fire is highly dangerous, particularly as the floors will be hosed down with water.

Stone, concrete or tiled floor   The floor should be one that can easily be washed down with a hosepipe and brush. A channel and drain to take away the surplus water is essential. If the floor is concrete, it is necessary to ensure that the rendering is carried out to a fine finish so that there is no likelihood of dust.

Cool, clean, working surfaces   The surfaces should be easily cleaned and scrubbed without the possibility of harbouring dirt. Traditional marble or slate are ideal if you can get it, but stainless steel or formica are very good. Some people have wooden slatted working surfaces so that water drains straight onto the floor; these are fine provided dirt does not build up on the inner edges.

Sink with hot and cold water   Cold water is needed for hosing down and for such operations as milk cooling, and hot water is essential for adequate cleaning.

Cooker The traditional dairy did not have a cooker in it, and the cheese was made in a heated vat which stood on the floor, but as most people are not likely to have such a vat, a cooker is a sensible alternative. It also provides a constant source of boiling water, although many people have found the small Burco boilers to be excellent for this purpose. The cooker or boiler should have the electric point high up, and ideally, should be on supports which keep it clear of the ground, and hence away from water. People with Agas or Rayburns in their kitchens will obviously prefer to utilize this heat for their cheesemaking, rather than pay for extra heating somewhere lese.

Clean, dust-free storage Most of the equipment used is adequately stored upside down on a working surface, but smaller items such as thermometers, acidmeters, rennet and so on are better kept out of the way in a cupboard.

Refrigerator Lactic starters and things with a limited life such as cream cheeses are best kept under refrigerated conditions. The refrigerator of course does not need to be in the dairy itself.

Safety factors The danger associated with floor electric fires in a room where there is a lot of water has already been emphasized. Heavy items such as a centrifugal cream separator will need to be bolted to a sturdy working surface, unless it has its own stand. Knives and any dairy chemicals should be stored well out of reach of children.

The kitchen as a dairy

For most people it is their kitchen which acts as a dairy and there is no reason why perfectly good dairying operations should not be carried out here. In many ways a kitchen is better than the traditional household dairy because it is equipped with electric light, power points and water heating facilities, and is probably a lot cleaner. Where a problem is likely to arise is in the fact that many houses are centrally heated and a kitchen is sometimes too warm. If you have a cool pantry it is better to utilize this for leaving milk to set or to store cheeses while they are ripening. There is also the fact that a kitchen is used for a number of activities where the traditional cool room was reserved only for dairying. It is important to make sure that anything to do with bread making or fruit preserving is not left lying around. The yeasts associated with these activities can adversely affect your cheese and yoghurt. It is obviously important to exclude pets from the kitchen while any dairying or cooking operations are taking place.

The conditions for dairying

Scrupulous cleanliness is essential. In many dairying activities cultures of bacteria are being utilized for a particular purpose such as the ripening of cream or the acidifying of milk. It is important to exclude those that are not desirable and which may have the opposite effect to the desired one. Milk is an excellent medium for the growth of many bacteria, good and bad.

Working surfaces should be thoroughly clean and all equipment and utensils sterilized before use. Sterilization means immersing or scalding in boiling water and it is here that the Burco boiler, referred to earlier, comes in useful. For sterilizing small quantities of equipment, an electric kettle to pour boiling water into a basin is perfectly adequate. A large saucepan of water will also suffice. On a larger scale, a purpose-made dairy or hypochlorite disinfectant is available from dairy suppliers. If hypochlorite is used, all equipment should be well rinsed because if any traces remain, it may have the effect of killing off the required bacteria in the milk products.

The best material for dairying utensils is stainless steel which is easily cleaned and

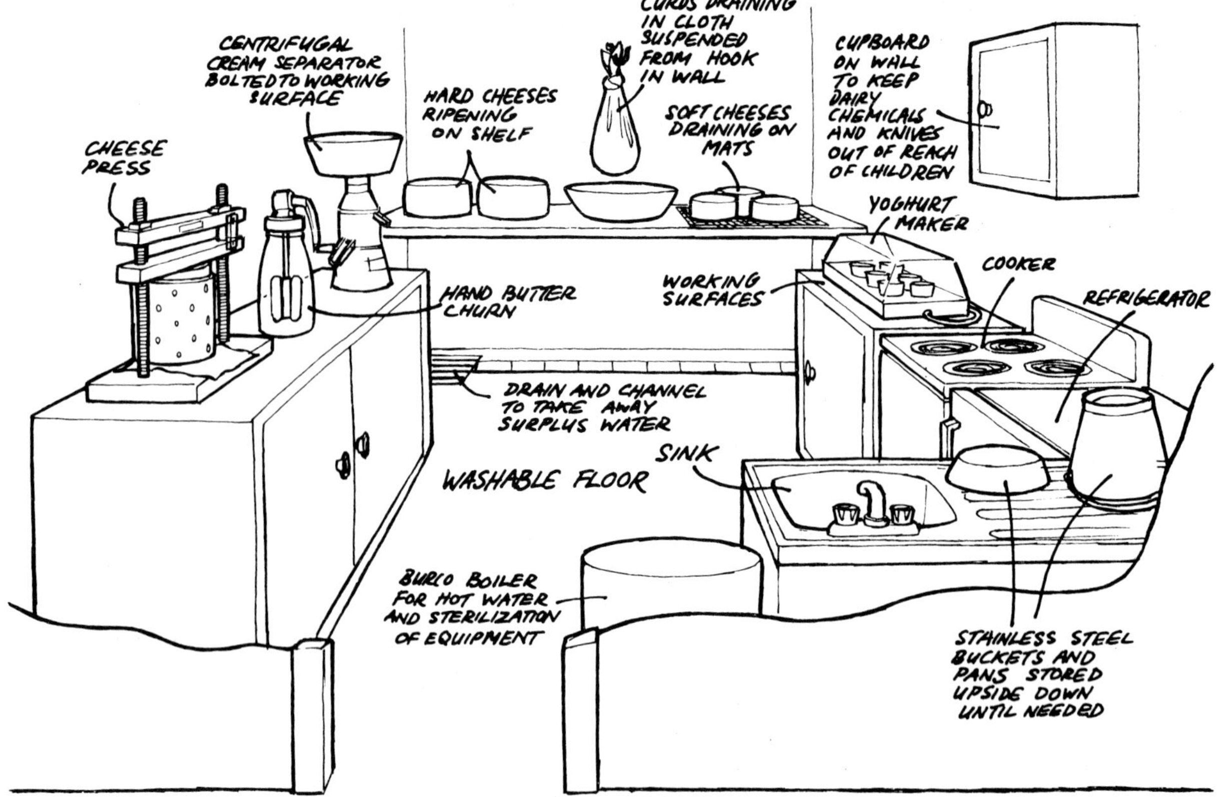

Fig. 9 *Interior of a small home dairy.*

resistant to scratches. Many of the traditional tools may be nice to look at in farm museums but they were diabolical to clean. Wood, pottery, enamel, tin, brass and copper have all been used in the past, but are not recommended for modern use. If you are thinking of buying such bygones in farm sales, think again. There are now many suppliers of small-scale dairying equipment such as cheese presses, cream separators, cheese moulds, small vats and buckets. Modern equipment is made of stainless steel, aluminium or high density plastic, all of which are far superior to the traditional materials.

3

MILK

Dairy animals have been kept for the provision of milk since the beginning of recorded history. Milk is one of the most complete foods available, containing proteins, fats, sugars, minerals and vitamins held in water. The composition varies slightly depending upon the breed of dairy animal, its genetic make-up (whether or not it comes from a good milking strain), and the period of its lactation. There is also a variation between the first, middle and last milking or 'stripping' with the highest concentration of fat being obtained at the end of the milking. Milk is an incredibly complex substance as will be appreciated by seeing how many things it contains. A breakdown of its composition is shown below.

Average composition of cows' milk

	per cent
Water	87.1
Fat	4.0
Casein	2.5
Albumin	0.7
Lactose	5.0
Ash	0.7

Breaking it down even further we find that fat is composed of ten substances – dioxystearin, olein, stearin, palmitin, myristin, laurin, caprin, caprylin, caproin and butyrin. Proteins are present as casein, albumin, lecithin, globulin and fibrin. Minerals are represented in the form of potassium oxide, sodium oxide, calcium oxide, iron oxide, sulphur trioxide, phosphorus pentoxide and chlorine. Lactose is the milk sugar and citric acid is also present in small quantities. Finally, vitamins, including A and D are present, although these are destroyed by boiling.

As referred to earlier, milk composition varies for a number of reasons, with the level of fat varying more than most of the other components. The relative fat levels of different milks are shown below, but it must be appreciated that it is a general guide only and does not take into account variations in different strains.

	Percentage of milk fat
Jersey cow	5.5
Guernsey	5.3
South Devon	4.5
Dairy Shorthorn	4.0
Ayrshire	3.8
Friesian	3.6
Holstein	3.3
Goat	3.5
Sheep	6.5
Ass	1.8
Human	3.8
Mare	1.09
Camel	2.9
Sow	4.5
Cat	3.3
Dog	9.5
Porpoise	45.8
Elephant	19.6

Types of milk

The following types of cow's milk are defined for the purposes of commercial distribution in the UK. There are similar designations in the USA, Australia, New Zealand and other parts of the world.

Untreated or raw milk   This is milk which has been filtered and cooled then bottled in containers with an identifying green top. It has a higher vitamin content than other milks.

Pasteurized milk   This is bottled in silver-topped bottles and has undergone heat treatment to destroy bacteria. About 20% of the vitamins are also destroyed.

Homogenized milk   Found in red-topped bottles, homogenized milk has been heat treated and subjected to pressure in order to break up the fat globules so that they are evenly distributed through the liquid. There is therefore no cream line at the top and the milk is easier to digest because of the small fat particles.

Sterilized milk   Milk which has been heated to 100°C (212°F), homogenized, then held at a high temperature after bottling and seal-ing. It destroys all the bacteria as well as the vitamins and the 'cooking' of the milk leads to slight caramelization of the lactose sugar, giving the milk a distinctive taste.

Ultra Heat Treated (UHT) milk   The com-mon name for this is 'long life' milk and it is usually sold in foil-lined waxed paper con-tainers. It has undergone a 'flash' heat treat-ment of being subjected to temperatures between 135°C and 150°C (275°F–300°F) for a few seconds only.

Channel Islands milk   This is milk with a legal minimum level of 4% fat and is pack-aged in gold-topped bottles where it has been pasteurized or a green top with a gold stripe if it is raw. South Devon milk is also sold as 'gold-top'.

Dried milk   There are considerable sales of dried cow's milk utilizing the skimmed milk left over from the cream and butter industries. It is popular with those on a diet and is useful for camping. Dried goat's milk is also available and New Zealand has pro-duced most of Britain's imports to date. In recent years however, a goat cooperative based in Wales has begun to produce dried goat's milk for use in hospitals and for those with allergy or digestive conditions.

Evaporated milk   This is sold in cans and is milk which has been heated in a vacuum until it is reduced to half of its original volume.

Condensed milk   The milk is heated and reduced to two fifths of its volume and mixed with sugar. It is marketed in cans.

Cows' milk is sold in Britain through the Milk Marketing Board who operate a highly efficient distribution system. Britain is one of the few countries in the world which can claim to be able to deliver fresh milk to the doorstep of every household in the land each morning. Goats' milk has no central distribution system and the bulk of it is provided by small-scale producers. Dis-tribution tends, therefore, to be more haphazard, with supplies reaching retailers at more infrequent intervals rather than every day. This means that a large pro-portion of goats' milk is being bought by customers when it is several days old, a situation which would be unthinkable with cows' milk, unless it had been given special treatment to prolong its shelf life. This situation more than anything else has led to the common belief that goats' milk tastes 'goaty'. If cows' milk which had not been pasteurized was sold when it was several days old, it too would taste peculiar. In the USA, Australia and New Zealand, distribu-tion of tinned goats' milk is fairly common, and, on recent visits to California and Aus-tralia, I was able to buy tinned goats' milk which had no taints whatever. In Britain it is only possible to buy frozen goats' milk as an alternative to fresh, although, as referred to earlier, dried milk is available in limited quantities.

Milk treatment in the dairy

Once milk has been taken from the dairy animal there are various processes through which it must pass. These processes are the same regardless of the scale of operations; it is merely the equipment which varies.

Filtering

The milk is first filtered in order to remove any foreign bodies such as dust particles or hairs from the animal. In a commercial unit the milk would be piped directly from the milking unit to a cool room equipped with a filtration unit. On a smaller scale, a filter unit is commonly used which is equipped with disposable filter discs. On an even smaller scale, where, for example, only one goat is kept, a sterilized kitchen strainer with a filter disc in it can be used.

Cooling

Unless the milk is to be pasteurized it needs rapid cooling to prevent bacteria multiplying. On a large scale, this process is virtually automatic with rapid cooling facilities in the milk tank. On a smaller scale an in-churn cooler is frequently used. This is where a metal container, through which cold water flows, is inserted in the churn holding the milk. Where only a small amount of milk is involved, the easiest way of cooling is to place it in a metal milk can with lid and stand it in a sink of cold, running water.

Pasteurization

On a small, domestic scale it will not be necessary to pasteurize milk unless the aim is to make the milk last longer. If the weather is hot and humid it is sometimes worth doing. For small amounts, it is achieved by heating the milk in a large saucepan until a temperature of 66°C (150°F) is reached. (A purpose-made dairy thermometer has the pasteurization temperature clearly marked.) I treat some of my goat's milk in this way because heating makes the cream rise to the surface, from where it can be skimmed. The skimmed milk is then heated to pasteurization temperature and transferred to sterilized bottles which are capped and then refrigerated. Ordinary milk bottles are not normally available for use in this way because they are the property of dairies who supply the public with bottled milk on the understanding that the bottles are returned. Lemonade bottles are perfectly adequate and if caps are not available, a small amount of cooking foil can be placed on the tops. Where pasteurization is necessary on a slightly larger scale than this, small units catering for several gallons are available from specialist suppliers.

Most cows' milk sold in Britain has been pasteurized, but goats' milk is normally sold raw.

Storing

Milk should be kept under refrigeration when it will last for several days, particularly if it has been pasteurized. Where storage for longer periods is necessary, it will need a different preservation technique. Canning and the ultra heat treatment (UHT) referred to earlier are beyond the scope of the small scale producer and freezing is the only possibility. Cow's milk does not freeze satisfactorily because the cream does not reconstitute properly afterwards. This is not to say that it will not freeze. It will do so as long as you do not mind the particles of cream that float on the surface when it thaws. Goats' milk freezes well, although there is also a slight deterioration in the cream. Three months is the limit for which it should be kept under refrigeration. Some goatkeepers who sell milk claim that better storage is achieved with a commercial deep freezer than with a domestic one where the temperature is not as low.

Packaging and selling milk

Cows' milk may not be sold in Britain without being registered with the Milk Marketing Board. Such an enterprise would not be viable unless a comparatively large herd was kept. Before registration, the provision of adequate buildings and conditions are required and the water supply is tested for purity, in order to meet the requirements of the dairy regulations. Anyone who is interested in this aspect should contact initially the dairying officer of the local department of agriculture. Similar requirements are in force in the United States, Canada, Australia and New Zealand whose respective agriculture departments will provide information on request. The addresses are given in the reference section.

Goats' milk can be sold without a licence, although the normal health regulations which are applicable to all foods will apply here. In other words, if goats' milk is sold and is found to contain impurities which may be detrimental to health, it will be in breach of regulations and the vendor is liable to be prosecuted. Most local authorities will test the milk being sold and may wish to see the dairying conditions under which it is being produced.

There are two ways in which goats' milk is packaged for sale in Britain: in waxed paper cartons sealed with metal clips; and in plastic bags. The former are used for fresh, liquid milk, while the latter are used for the sale of frozen milk. Both containers are available, pre-printed with an illustration and description of the contents. It is legally required to state the quantity and describe the contents on the package. In the United States, Australia and New Zealand, goats' milk is frequently canned, but this is usually carried out at the registered premises of a marketing cooperative which arranges distribution.

Recipes using milk

These recipes are appropriate for anyone who has milk to use, whether it is home-produced milk or that provided by the milkman.

Milk shakes

Milk shakes are easily prepared in a blender. For every cup of milk use one teaspoonful of ice cream of the appropriate flavour. Where ice cream is not used, try any of the following: two blended fresh strawberries; 2 teaspoonfuls drinking chocolate; 1 teaspoonful instant coffee with sugar to taste; a quarter of a blended fresh orange. Vanilla ice cream can be used with any of the above ingredients for a creamier milk shake.

Treacle posset

This is a traditional drink for cold winter evenings. Stir a teaspoonful of black treacle into a cup of hot milk.

Home-made custard

500 ml (1 pint) (2½ cups) fresh milk
1 level tablespoon (1½ tablespoons) cornflour
1 dessertspoon sugar
1 egg

Blend the cornflour and sugar with a little of the milk in a bowl. Beat the egg and stir into the mixture. Heat the milk separately and when it is about to boil pour into the bowl, stirring continuously. When it is well mixed, transfer the mixture to the saucepan and heat carefully, stirring all the time until it thickens.

Rice pudding

100 g (4 oz) (½ cup) short grain rice
100 g (4 oz) (½ cup) sugar
1 litre (2 pints) (5 cups) milk
Pinch of salt

Put the rice, salt and sugar in a heatproof dish and add the milk, stirring well. Cook in a low oven at 150°C (330°F) Gas Mark 3 for three hours. Give it a stir every now and again to stop the grains settling in the bottom while it is cooking. It is better to make this when the oven is also being used for something else in order to make the best use of energy. Traditionally rice pudding was made in an oven which was cooling down after the bread and pastries had been cooked. Using a pressure cooker will shorten the cooking period considerably.

Junket
500 ml (1 pint) (2½ cups) milk
1 tablespoon (1½ tablespoons) sugar
1 teaspoon junket rennet or 1 rennet tablet

Warm the milk to blood heat when it will feel warm to the touch. Add the rennet and sugar and stir well. Leave covered in a warm place until it has set. Avoid disturbing it until it has set. Junket tablets are normally flavoured already and may be purchased with the appropriate flavour. It is best eaten warm rather than refrigerated.

Note: Use either metric, *or* imperial, *or* American measures in each recipe.

4

YOGHURT

Yoghurt is milk which has been coagulated into soft curds by the action of one of the lactobacillus organisms. The usual ones are *lactobacillus bulgaricus*, *lactobacillus acidophilis* and *streptococcus thermophilus*. These micro-organisms feed on the milk sugar lactose, producing lactic acid which, in turn, acts on casein, a milk protein, bringing about curdling. The lactic acid gives the yoghurt its characteristic taste and the longer it is left, the more acidic it becomes. It can be made from good quality cows', goats' or ewes' milk and in the past has also been produced from the milk of mares and asses.

No-one can be certain when or where yoghurt first appeared, but it is likely that it was the accidental, natural souring of milk which led to its discovery. What is certain is that it has been a traditional dish in the Caucasus, Bulgaria, Greece, Turkey and other areas of Europe for many generations. Myths abound about the beneficial effects of yoghurt and its contribution to the longevity of some inhabitants of the Caucasus: many people claim that lactic acid actively encourages the body to resist disease. Whatever the facts may be, yoghurt is a popular and nourishing dish which many people enjoy.

Making yoghurt at home

Making yoghurt at home is not difficult as long as proper regard is paid to the provision of hygienic conditions. All equipment and utensils should be sterilized in boiling water so that there are no harmful bacteria around. The equipment needed is simple, a saucepan to heat milk, a thermometer to ensure that the correct temperatures are achieved and a protected container to hold the yoghurt while it is incubating. The latter could be a thermos flask, a small shop-bought unit or a home-made incubator utilizing a plant propagator such as the one shown in figure 10. Hay boxes have also been used to good effect.

The principle of yoghurt making is to first kill off unwanted bacteria by heating the milk then, after cooling to the optimum temperature of 43°C (110°F), introducing a 'starter' of the correct lactic-acid-producing bacteria. Plain yoghurt bought in the shop can provide a culture of the necessary bacteria, but there is no guarantee that this will be strong enough. It can sometimes work well, and at other times, nothing happens. For consistently reliable results, it is safer to use a commercial starter such as a freeze-dried lactic ferment culture, or a liquid one purchased from a reliable source. Some people believe that it is necessary to add powdered milk in order to thicken yoghurt. This is not true. Commercially it is added because skimmed milk as a by-product of cream production is being used, rather than whole milk. If milk, with a normal butterfat content is used, the optimum conditions are provided and a viable starter is added, there should be no problem producing firm yoghurt. Full cream Channel Islands milk or ewe's milk will produce a thick yoghurt often with a

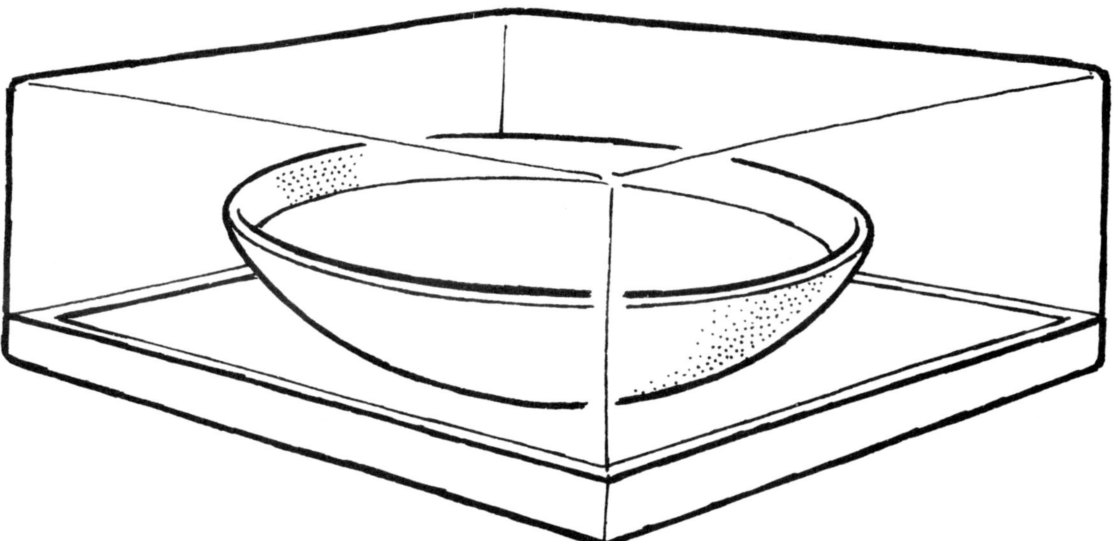

Fig. 10 *A plant propagator with a heated base plate and rigid plastic cover can be used for incubating the yoghurt.*

delicious crust on the top. Visitors to Greece may have sampled the ewe's milk 'yaorti' with its rich cream surface.

Dairy suppliers sell a strong culture, freeze dried and packeted in foil. It is date-stamped to indicate when it should be used by (usually six months), and until used it needs to be stored in the deep freezer. When needed for yoghurt making it can be propagated in exactly the same way as cheese starter (see page 51), the only difference being that the incubation temperature needs to be higher i.e. 43°C (110°F) for 3–4 hours. This can be achieved by filling a wide-necked Thermos flask with water at this temperature, and inserting the incubating bottle.

The basic recipe for making yoghurt
500 ml (1 pint) (2½ cups) full cream milk
5 ml (1 teaspoon) (1 heaped teaspoon) starter
 or plain yoghurt
(If skimmed milk is used, add 20 ml (1 tablespoon) milk powder.)

Heat the milk to 82° (180°F) or just below boiling point. Cool the milk to 43°C (110°F) when it will feel hot but bearable to the finger. Pour most of the milk into a thermos flask which has been previously rinsed with boiling water. Blend 1 teaspoonful (1 heaped teaspoon) plain yoghurt or prepared lactic culture starter with the remaining milk and add to the thermos flask. Replace the lid of the thermos firmly, give it a shake and leave undisturbed until a firm curd has formed. Remove the lid and leave the yoghurt in the refrigerator to cool and firm. It is then ready to eat plain, with honey added, or mixed with fruit.

Producing flavoured yoghurts

It is not difficult to produce flavoured yoghurts, as long as several basic rules are observed. The first is that adding anything before the period of incubation is best avoided unless absolutely essential. Wait until the yoghurt has been produced before gently mixing in the desired flavouring. Added ingredients tend to fall to the bottom, although this is obviously more likely with solid additions such as fruit. Some powdered ingredients such as chocolate powder can be stirred in before incubation. The second point worth remembering is not to add too much of anything that is likely to dominate the flavour of the yoghurt. It is difficult to generalize about this and per-

1 Heat milk to 82°C (180°F) and cool to 43°C (110°F)

2 Put in thermos flask

3 Blend in starter and add to thermos

4 Leave undisturbed for curd to form

Fig. 11 *Step-by-step yoghurt making*

sonal experimentation is the best approach. The yoghurt will gradually become more acidic as it matures, so avoid accelerating this by adding too many acid ingredients. Finally, adding anything must be done gently so as to avoid breaking up the curd and producing a thin, runny yoghurt. Incorporate the flavouring with the absolute minimum of stirring. If there is a little separation, you can cheat by pouring off some of the liquid whey. Putting the flavoured yoghurt in the refrigerator for a few hours before use will firm it up again. There is a wide range of flavourings and it is a matter of personal taste, but some of the more popular ones are: honey, whole or chopped fresh fruit and tinned fruit (but avoid too much juice).

Some problems

If, despite all your efforts, the yoghurt does not firm, it could be for one of the following reasons:

Using a starter which is too weak Buy a fresh one from the wholefood shop, or use a commercial lactic culture starter.

Adding the starter when the milk is too hot Overheating kills off the bacteria. Use a dairy thermometer.

Detergent or sterilizer tainting milk The effect of both of these is to kill off the yoghurt bacteria. Make sure that all utensils are thoroughly rinsed in scalding water.

The temperature is too low This is where reliance on an airing cupboard or similar situation lets you down. The temperature of 43°C (110°F) needs to be maintained for several hours.

The milk has too low a butterfat content This is more likely to occur with some goats' milk if the quality of the milk is poor, another reason for ensuring that only good stock is kept. Buy pedigree goats that come from milk recorded stock. Anglo-Nubians with their naturally high butterfat milk are a good choice.

Antibiotics in the milk   When all the above possibilities are eliminated, there is one that remains – the possibility that the milk is affected by antibiotics used to treat the dairy animal for a condition such as mastitis. If the milk is from the milkman ask your local health inspector to arrange an analysis test. It is strictly illegal to sell milk affected in this way, but it is unofficially recognized that it does happen, for mastitis is an indigenous condition in many heavy-yielding dairy herds.

If it is your own cow or goat involved, there should be at least three clear days allowed before the milk is used, after penicillin is administered.

Small-scale commercial yoghurt production

Yoghurt is sold in every supermarket in the land, but a high proportion of it bears little resemblance to the genuine article because it has so many additives such as artificial flavourings, colourings and stabilizers. In recent years there has been something of a reaction against commercial adulteration of foods and the small-scale producer of yoghurt containing only natural ingredients has found a ready market.

Only a few years ago there was a shortage of equipment for the small producer but the situation has now changed, with a considerable range being available. There are basically two types of commercial yoghurt producer: the vat-type and the cabinet type. The former has a capacity of 30 litres (5 gallons) and milk is heated in an internal and lidded stainless steel bucket, then cooled by a system of draining, refilling with water and agitating the product. Commercial starter is added and the solid state electronic temperature control set. After approximately 3½ hours the yoghurt is ready for cooling and flavouring as necessary. The cabinet type producer has a stain-

Fig. 12 *A small yoghurt kit*

less steel interior with removable shelves and an adjustable temperature control (Fig. 12). Individual pots are placed on the shelves for incubation.

Packaging used to be a problem for small producers, but, again, manufacturers have made equipment available. Plastic pots with snap-on lids can be used or a small hand-operated capping machine for heat-sealing foil tops is available (Fig. 13). These tops are available plain or pre-printed with the illustration of the fruit contained in the yoghurt. It is simply a matter of placing the filled yoghurt carton in the supporting base container, sliding it under the capping head which is furnished with a foil cap and then pulling the handle. The cap is sealed by a combination of heat and pressure.

Commercial yoghurt production should take place in a properly equipped cool room where hygienic conditions can be maintained. Sterilization techniques for equipment would include either the use of boiling water or a proprietary dairy sterilizer such as hypochlorite solution. All traces should be removed in case it kills the yoghurt starter culture. A licence will be necessary for the production of cows' milk products and the best place to obtain help and information about this is the dairy husbandry advisory officer at the local branch of the department of agriculture. The address and telephone number is in the local telephone directory or Yellow Pages directory but regional addresses are also given in the reference section of this book. With goats' milk products a licence is not

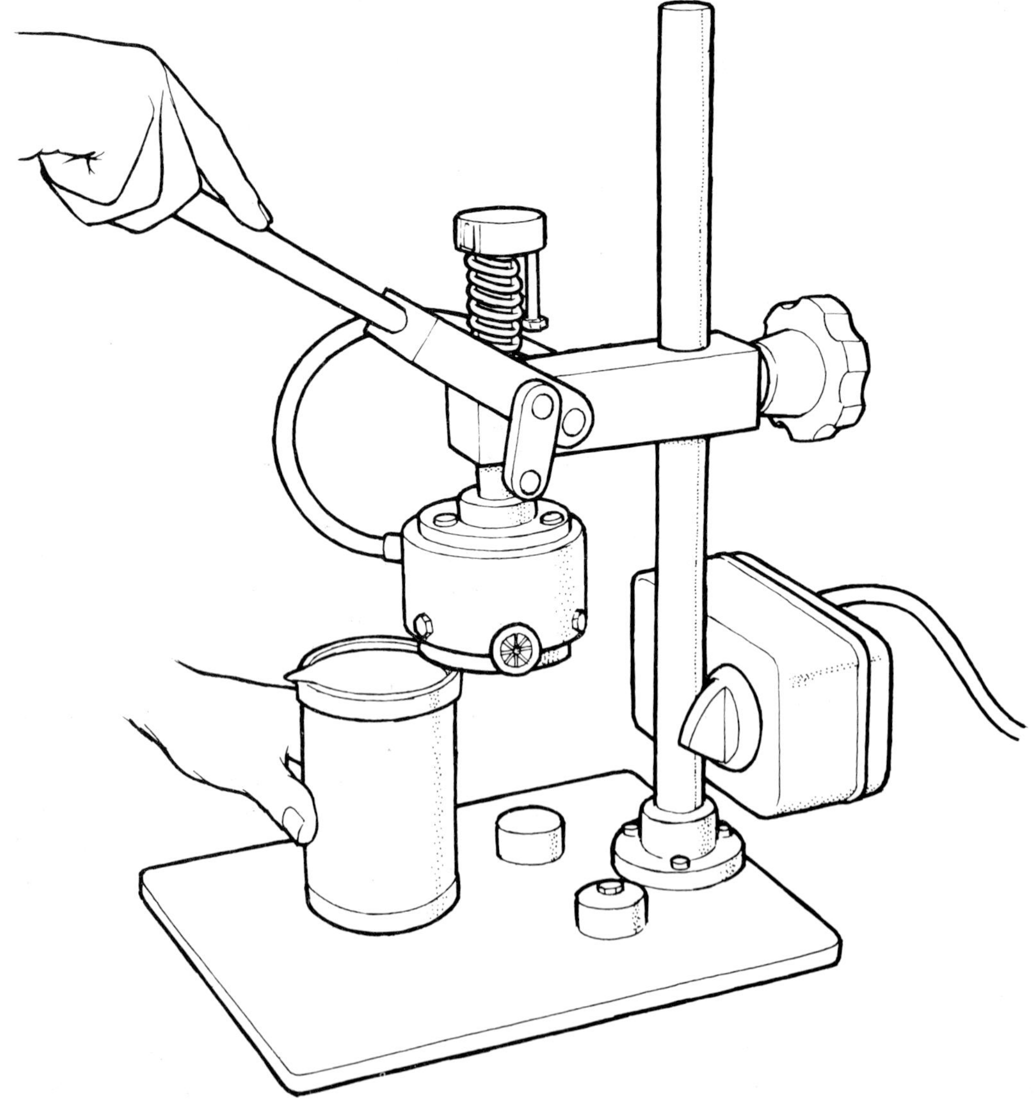

Fig. 13 *Method of capping yoghurt pots with foil for the small, commercial home production of yoghurt*

necessary, but the same hygienic standards that relate to the production of all foods will apply. All yoghurt offered for sale must have the quantity marked on the container, as well as having the ingredients listed. An example of the latter might be: 'Dairy goat's milk: live yoghurt culture: fresh oranges'.

Other forms of yoghurt

Kefir

Kefir originated in the Balkans and is a traditional drink of fermented milk produced by the addition of kefir grains. These grains consist of dried milk solids containing lactose-fermenting yeasts as well as lactobacilli. They look rather like brown, wheat grains until added to milk when they swell, turn white and grow into small 'plants' rather like cauliflower florets. For this reason they are often referred to as 'yoghurt plants'.

Milk is heated, to destroy bacteria, then cooled to 43°C (110°F) before the grains are sprinkled on and stirred. If left overnight in warm conditions, the yoghurt will form and the grains themselves are then removed and dried until required for further use. They will last for at least a year, and possibly longer if looked after. If the kefir drink, rather than the firm yoghurt is required, stir it up well and leave for a further 24 hours until the fermentation and slight alcoholic content is reached.

Koumiss
Koumiss originated in Russia and was made either from a combination of cow's and mare's milk, or from mare's milk alone. These days it is made from cow's milk only. It is fermented by the addition of lactose-fermenting yeasts and curdled by the action of *lactobacillus bulgaricus* until the product looks like a greyish white, slightly fizzy liquid.

Cultured buttermilk
Buttermilk was originally the liquid left behind after milk was churned to produce butter. If left it will gradually become acidic as a result of natural souring. It is a traditional drink in Wales, and babies were weaned onto it. My childhood memories of Wales include the large earthenware crock with the round slate lid which we used to store the buttermilk, and the large ladle which hung nearby, to scoop out a cool, refreshing drink whenever it was needed. Cultured buttermilk is a different product and is made from skimmed or partially skimmed pasteurized milk which is innoculated with a lactic-acid producing culture or starter and which has a longer shelf life than the traditional buttermilk which is now officially referred to as acidulated buttermilk.

There is a wide variety of yoghurt-type drinks in different parts of the world. The variations may include differences in butterfat or milk protein content or degree of acidity and fermentation. Some examples are as follows:

Scandinavia	*USSR*	*Egypt*
fuli	kefir	leben
puma	koumiss	*India*
filmjölk	araka	dhai
lakstofil	skorup	*Chile*
gräddfil	kos	grusavin
längfil	*Yugoslavia*	*Sardinia*
pitkäpiimä	kaimae	gioddu
skyr	*Switzerland*	
villi	fru-fru	
surmelk	*Hungary*	
	tarko	

Vegetarian yoghurt
There are not many vegetarians who are so vegetarian as to not eat ordinary yoghurt but nonetheless it may be worth knowing that yoghurt can be made from soya milk. This plant 'milk' is available from most health food shops and it is claimed that a passable yoghurt can be produced by curdling it with a lactic acid culture or keffir grains referred to earlier. I have never tried this recipe, however, and I cannot therefore recommend it, but it sounds worth experimenting with.

Storing yoghurt

Yoghurt incubated at 43°C (110°F) will be ready at any time between 2–10 hours, depending on the quantity of milk and type of starter. Once curdled, it should be stored in a refrigerator and is at its best eaten after two days. After this it will become increasingly acidic but can be eaten at any time up to ten days. If there is any left after this time, it is best to discard it. While in the refrigerator, it should not be stored near strong tasting or flavoured foods which may impart taints to it.

Cooking with yoghurt

The best type of yoghurt for cooking is that which has a thick consistency. In India and parts of Greece and Turkey the curds are often strained through muslin so that the finished product, without its whey, looks rather like curd cheese. Sometimes the custom is to boil the original milk until a proportion of the water is evaporated, leaving a thicker consistency. It is then cooled and incubated with culture so that a thick yoghurt is produced. This is often the practice of goatkeepers whose goats are producing thin milk with a low butterfat content.

Where a recipe calls for the addition of yoghurt and that to hand is of a thin consistency it will need to be stabilized otherwise it will separate, leaving small particles in the mixture. This is simply a matter of beating 1 litre (2 pints) (5 cups) of yoghurt until it becomes runny, and then adding to it the beaten egg white of one egg mixed to a paste with 20 ml (1 tablespoonful) of cornflour and 20 ml (1 tablespoonful) of milk or water. Bring slowly to the boil, then simmer over a very low heat until it resembles thick cream. It will store in the refrigerator for up to two weeks.

Recipes using yoghurt

Cole Slaw
227 ml (8 fl oz) (1 cup) of yoghurt, 40 ml (2 tablespoonfuls) (3 tablespoons) of lemon juice or wine vinegar, 450 gm (1 lb) (2⅔ cups) shredded white cabbage, 114 ml (4 fl oz) (½ cup) mayonnaise, 1 grated carrot, salt to taste. Shred the cabbage and grate the carrot, then blend the yoghurt, salt and lemon juice. Mix thoroughly.

Potato Salad
Yoghurt can be mixed with any left-over potatoes to produce an interesting addition to a cold lunch. It is particularly nice with chopped hard-boiled eggs, salad onions and a sprinkling of parsley.

Soups
Many soups are improved by the addition of a tablespoonful of yoghurt just before serving.

Salad dressing
Mix one cup of plain yoghurt with 5 ml (1 teaspoon) of prepared mustard, ½ teaspoon salt, 1 crushed garlic clove and a pinch of paprika. When all the ingredients are thoroughly mixed, beat in 2 tablespoons (3 tablespoons) of lemon juice, adding a little at a time. Chill before use.

Apple yoghurt mousse
Bramley apples are the best to use for this recipe because they purée so well. Prepare the apples and stew with just enough water to stop them burning. Add sugar or honey to taste, then to every 500 ml (1 pint) (1½ cups) of apple purée, add 300 ml (½ pint) (1¼ cups) of thick, plain yoghurt. Leave in the refrigerator until absolutely cold and, meanwhile, beat 4 egg whites adding 50 gm (2 oz) (4 tablespoons) of caster sugar when stiff. Fold it into the apple/yoghurt mixture and put into the refrigerator. Just before serving, sprinkle a little powdered cinnamon on the surface. Recipes for yoghurt cheesecake and yoghurt cheese are given in the appropriate sections.

5

CREAM

Cream is made up of small globules of fat composed of a number of different substances each of which has its own distinctive characteristic. For example, the olein content of fat gives fresh cream its mellow taste, while stearin binds the whole together.

The cream content of milk varies depending on the time of year and also on the breed of dairy animal. Channel Island cows and those of North and South Devon have milk with a higher fat content and the globules are larger and more yellow than in other breeds. Anglo-Nubian goats have a higher butterfat or cream content to their milk, than most other breeds of goats. Dairy ewes tend to have a relatively high concentration of fat in the milk, although the content varies depending on the stage of their lactation.

The following are the different types of cream which are commercially available and which originate from cow's milk. Cream from the milk of other stock is not generally available. The difference between them is in the process they have undergone and in the degree of butterfat they contain; this can be regulated by adjusting the speed of the centrifugal separator when the cream is separated from the rest of the milk.

Clotted cream	55% butterfat
Double cream	48% butterfat
Whipping cream	35% butterfat
Sterilized cream	23% butterfat
Single cream	18% butterfat
Half cream	12% butterfat

Cream separation

Globules of fat are naturally lighter than the liquid milk medium in which they are contained, and if left to settle, will rise to the surface. Even goats' milk which is naturally homogenized, where the fat globules are small and evenly distributed throughout the milk, will behave in this way, although to a lesser degree than cow's milk. If the goats' milk is carefully heated, however, more of the cream will surface.

Cream may be separated in the following ways:

With a hand skimmer Leave the milk to settle for 12–24 hours in wide, shallow skimming pans until the cream has surfaced. Make sure that there are no strong-smelling substances stored nearby otherwise they will taint the milk. Skim the cream layer off the surface with a hand cream skimmer, ladle, or thin china saucer. The advantage of the skimmer is that it has perforations which allow the surplus liquid to drain away.

With this method it is not possible to obtain all the cream from the milk and no matter how careful you are, between 0.5 and 0.7% of fat will be left behind in the skimmed milk. With goat's milk of course, the percentage left behind is far higher.

Hand skimming was the traditional method of cream separation and the skimmers were referred to as 'fleeting' dishes. They were solid, saucer-like utensils made of horn or wood, and later of perforated tin.

1 Cream setting pan with hand skimmer

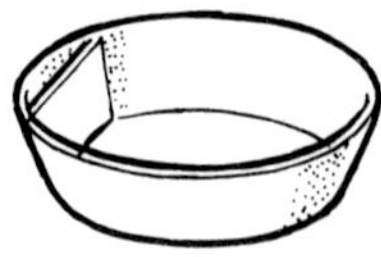

2 Combined setting/skimming pan with detachable lip

3 Centrifugal cream separator

Fig. 14 *Methods of cream separation*

The Castle museum at York has an interesting collection of early dairying utensils, including various 'fleeting' dishes.

Combined skimmer/setter   This method is essentially the same except that the cream is not skimmed manually off the surface. The bowl is tipped to allow skimmed milk to drain away while a special lip retains the cream. Again, there are many examples of traditional wooden and ceramic utensils with this feature, as well as modern pans made of aluminium or stainless steel.

Using a centrifugal separator   This is by far the most efficient method of separating, which can leave the separated milk with as little as 0.1% of fat. The principle on which the separator works is as follows: Milk is composed of particles of different densities. When it is rotated at speed within a container, the lighter, fatty particles will stay in the central axis, while the heavier particles of separated milk fly outwards. Centrifugal force increases as the square of the number of revolutions, but decreases in relation to the diameter of the bowl. For best results there are three factors to be taken into consideration – an even flow of milk, an even speed, and an even temperature. The best time to separate milk is immediately after milking, while the milk is still warm and if for any reason this is not possible, the milk will need to be warmed to 40°C (104°F) before being put through the machine.

The cream outlet or 'cream screw' together with the pressure of milk flowing into the bowl determines the density of the cream. It is therefore advisable to check whether the outlet is suitable for your milk, particularly if you have goats' milk where the fat globules are smaller. It is essential to dismantle and thoroughly clean the separator immediately after use, and this can be a time-consuming business. The first centrifugal separator appears to have been invented in the nineteenth century and was hand-cranked. Hand models are still available, although electrically powered ones are obviously more appropriate if a large volume of milk is to be separated. They require scrupulous cleaning after use.

By siphoning   This method involves using a plastic tube to siphon off the milk from under the cream, without disturbing the rest of the liquid. It is appropriate only for the very small scale producer, such as an urban dweller who may wish to use 'top of the milk' cream from his bottled supply of milk from the local dairy. It was quite common during the war when butter was rationed for people to separate cream in this way and to produce their own small quantities of butter.

Storage of cream

As soon as cream is separated, the aim is to cool it as quickly as possible, to a temperature of 2°C (35°F) when it will store for about two weeks. For longer periods it should be frozen by putting into plastic bags, removing the air and placing in the fast-freezing compartment of the deep freezer. It will keep for up to two months in this way, but you may experience a certain amount of 'grittiness' in the reconstituted product and whipping it is not as successful as with the fresh product. The commercially frozen cream now available in supermarkets is more successful than the home-frozen product, not only because it has a high proportion of butterfat in relation to water but it is also quick-frozen, a feature that domestic freezers do not possess.

If the cream is to be used for making butter, it should not be stored in the refrigerator, but allowed to ripen at a temperature of 10°C (50°F) for at least 24 hours. This allows time for a certain amount of natural acidity to develop so that the resulting butter has more flavour.

Pasteurizing cream

Cream will last longer if it is pasteurized to kill off unwanted bacteria, and, if you wish to make butter from goats' milk cream, it is worth pasteurizing it so that there is less possibility of taints in flavour developing. Place the cream in a heat resistant bowl placed over a saucepan and stir it while it is over the heat. Raise the temperature of the cream to 63°C (145°F) but do not allow it to go beyond this or it will develop a scalded flavour. A dairy thermometer will be necessary for this. Once this temperature has been reached, remove the bowl from the heat source and place in cold water, ensuring that no water splashes into the cream. Then place it in the refrigerator until required.

Clotted cream

The art of clotted cream making appears to have originated in Britain, in the West Country, and Devon, Cornwall and Somerset are still the areas where most supplies come from. It no doubt came into being as a process because someone noticed that cream which had been scalded kept longer. Its distinctive taste is due to slight caramelization of the milk sugar and the coagulation of some of the proteins in the heating process.

Many of the farms in the West Country of Great Britain used to have a cream scalder which was a covered vat containing hot water in which one or more scalding dishes containing unseparated milk were placed in fitted holes so that steam bathed the bottom of them. The milk was left to settle overnight. It was then heated to boiling point over fires before being poured into the vat where the temperature was sufficient to maintain 77°C–88°C (170°F–190°F) for 30–50 minutes. After this time, the scalding pans and cream were lifted off the vats and placed on cool shelves in the dairy for about 12 hours in summer or up to 24 hours in winter. After this, the thick crust was skimmed off, allowing the liquid to drain through the holes of the skimmer.

Modern methods of clotted cream production have changed very little, although these days, the scalding vat is electrically heated and it has been found that there is less risk of souring if the process is accelerated. In other words, only two inches of separated milk are put in the bottom of the pans with a thick layer of separated cream on the surface. This is left to settle for 2–3 hours, then scalded to a temperature of 85°C (185°F) for 30–40 minutes. Cooling is usually brought about by running cold water through the outer jacket of the vat and, as soon as the cream is cool, it is skimmed and packaged. It is more bland than the traditional product which had a nutty, caramel flavour. It is still possible to achieve this by making clotted cream at home – there is no particular mystique about it, though it is necessary to have milk with a high butterfat content.

Making clotted cream at home

Take cream which was previously skimmed from milk in a setting pan or that from a centrifugal separator, and put it in a small saucepan or heat-resistant pyrex dish. Place this in a saucepan containing hot water and place the whole on the heat source. Bring to the boil and simmer for twenty minutes so that the cream is scalded by the steam, and a golden crust forms on the top. Allow to cool overnight and when quite cold, carefully skim off the crust.

Well-produced clotted cream made from high butterfat cow's milk will be golden in colour with a granular texture and no thin cream at the bottom, an indication that scalding is incomplete. If overdone, the texture may be gritty and have a streaky appearance, or may seem slightly 'oily'.

The colour of cream

There are several misapprehensions about the colour of cream. One of the more common (at least amongst urban dwellers) is that cow's milk cream is more yellow in spring because the cows eat buttercups. This may appear to be a novel and cheap way of feeding dairy animals, but, in fact, buttercups are poisonous and are certainly not eaten. What is true is that when lush green grass is eaten, the cream is more yellow, which explains why winter milk is always paler. The milk of Jersey cattle has more carotene in it than that of other breeds so it is naturally more golden at any time of year. When cream is scalded for the production of clotted cream, as described above, the slight caramelization of the milk sugars gives it a more golden colour than usual. Goats' milk cream is pure white.

Sour cream

Sour cream originated amongst the Slavic nations of Central Europe and is still a favourite as a salad dressing or for adding to other foods. Cream with a butterfat content of at least 18% is required and after pasteurization it is innoculated with a culture of lactic acid-producing bacteria. It is incubated at a temperature of 22°C (70°F) for 12–14 hours until an acidity of 0.6% lactic acid is achieved. Purpose-made cultures of *streptococcus lactis* and *leuconostoc cremoris* are obtainable from specialist suppliers, but for home production, a little plain yoghurt or yoghurt starter will produce a similar effect.

Commercially, the cream is homogenized before the addition of the culture so that the redistribution of the fat particles has a thickening effect. For home production this is not crucial and if the cream is particularly thin, a few drops of rennet will thicken it. If you wish to have thick sour cream then use thicker cream to start with.

Ice cream

Real ice cream is exactly what it says it is – cream subjected to icy conditions. Commercial ice cream in Britain contains non-milk fat, a fact which has not escaped the eagle eye of the EEC commissioners who wish to outlaw the term on the basis of its inaccuracy. It is not necessary to have an ice-cream maker, although it helps. The old ones with the crank handles were nice but often time-consuming to use. There are modern ones on the market and these are quick and efficient. Here is how to make it without a machine – although you will need either a deep freezer or an ice-making compartment in a refrigerator.

Step-by-step ice cream making

1 Unless you have a deep freeze, set the freezing compartment of your refrigerator to its lowest setting.

2 Whisk the mixture in a bowl, either with

Fig. 15 *Making ice cream*

Whisk the mixture

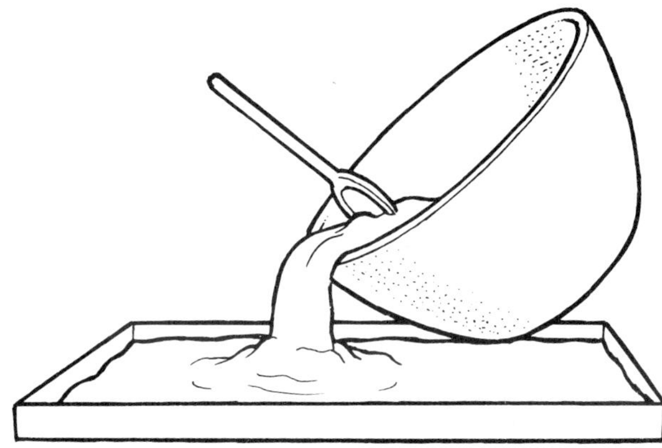

Pour into shallow trays

a hand whisk or mixer. (Refer to individual recipes for details of mixture.)

3   Pour the mixture into shallow trays.

4   Freeze until slushy.

5   Pour into bowl and whisk again to prevent large ice crystals forming.

6   Pour mixture back into shallow trays and return to freezing compartment until frozen.

7   Before use, remove from freezing compartment and place in ordinary refrigerator to soften for half an hour.

Here are some recipes which have all been tested:

Vanilla ice cream
2 separated eggs
50 gm (2 oz) (scant ¹/₂ cup) icing sugar
150 ml (¹/₄ pint) (⁵/₈ cup) double cream (heavy cream)
150 ml (¹/₄ pint) (⁵/₈ cup) milk
few drops vanilla essence

Beat the egg yolks, sugar and vanilla essence in a bowl. Bring the milk to boil, pour onto the mixture, stirring continuously, then leave to cool. Meanwhile whisk the egg whites until stiff and lightly whip the cream. Fold them into the egg yolk/milk mixture which by now should be quite cold. Follow the step-by-step directions as given above.

Chocolate ice cream
2 beaten eggs
85 gm (3 oz) (6 tablespoons) sugar
300 ml (¹/₂ pint) (1¹/₄ cups) milk
50 gm (2 oz) (scant ¹/₂ cup) cocoa
150 ml (¹/₄ pint) (⁵/₈ cup) water
150 ml (¹/₄ pint) (⁵/₈ cup) single cream (light cream)
150 ml (¹/₄ pint) (⁵/₈ cup) double cream (heavy cream)

Add the beaten eggs to milk and sugar and heat to boiling point, stirring all the time. Place cocoa in a pan, blend in water and stir over a gentle heat until it thickens. Pour it into the custard and leave until cold. Whip the two creams together until stiff, then stir them into the cold custard mixture. Follow the general directions given.

Tutti-Frutti ice cream
2 separated eggs
50 gm (2 oz) (scant ¹/₄ cup) chopped glacé cherries
50 gm (2 oz) (scant ¹/₄ cup) chopped mixed peel
50 gm (2 oz) (scant ¹/₄ cup) sultanas
150 ml (¹/₄ pint) (⁵/₈ cup) double cream (heavy cream)

Mix together the cherries, peel, sultanas and egg yolks. Whip the cream until stiff and fold into the fruit mixture. Continue as before.

Jamaican ice cream
2 beaten eggs
50 gm (2 oz) (4 tablespoons) sugar
2 large bananas
*20 ml (1 tablespoon) (1½ tablespoons) lemon
　juice*
300 ml (½ pint) (1¼ cups) milk
150 ml (¼ pint) (⅝ cup) double cream
50 gm (2 oz) (½ cup) chopped walnuts

Heat milk and sugar to boiling point then
pour over the beaten eggs in a bowl. Stir
and leave until cold. Mash the bananas and
lemon juice and lightly whip the cream,
then stir both into the custard. Continue as
before.

Cornflour custard ice cream
2 beaten eggs
85 gm (3 oz) (6 tablespoons) sugar
1 litre (2 pints) (5 cups) milk
*30 ml (1 tablespoon) (1½ tablespoons) corn-
　flour*

Blend the cornflour with a little cold milk,
and bring the rest of the milk to boil. Add
the blended cornflour and stir until it thickens. Whisk in the sugar and leave to cool.
Beat the eggs and add to the cold mixture.
Heat to just below boiling point, stirring
continuously, then cool.
The following recipes for which I am indebted to Quainton Dairy gives more
variations on the basic method. For really
rich, exotic and fattening ice creams, equal
quantities of single and double cream can
also be incorporated. They are not for those
on a diet, although it is worth remembering
the principle 'All things in moderation, including moderation'.

Maple syrup and walnut ice cream
*120 ml (6 tablespoons) (9 tablespoons) maple
　syrup*
50 gm (2 oz) (½ cup) chopped walnuts

Add the syrup to the cooled custard, then
continue with the basic recipe. When the
ice cream is partially frozen add the
chopped nuts. Return to freezer.

Ginger ice cream
*120 ml (6 tablespoons) (9 tablespoons) ginger
　syrup*
85 gm (3 oz) (1½ cups) preserved ginger

Method as above. Add chopped ginger
when ice cream is half frozen.

Peppermint and chocolate chips ice cream
2.5 ml (½ teaspoon) peppermint essence
50 gm (2 oz) (½ cup) plain chocolate

Stir essence into cooled custard. Add
coarsely grated chocolate and return to
freezer.

Coffee ice cream
*120 ml (6 tablespoons) (9 tablespoons) very
　strong black coffee, or coffee essence*

Add coffee to the hot custard. Cool and
continue with recipe.

Rich chocolate ice cream
110 gm (4 oz) (1 cup) plain chocolate

Grate the chocolate coarsely into the milk
for the custard. Bring slowly to the boil.
Continue with basic recipe.

Vanilla pod ice cream
1 vanilla pod

Split the vanilla pod in half length-ways.
Add to milk for the custard. Scald milk then
leave for 15 minutes to infuse. Take out the
vanilla pod. Continue with basic recipe.
(NB vanilla pod may be used again, rinse,
dry thoroughly and store in screw top jar.)

Caramel ice cream
110 gm (4 oz) (1 cup) caster sugar
120 ml (6 tablespoons) (9 tablespoons) water
*An extra 60 ml (3 tablespoons) (4½ table-
　spoons) water*

Put the sugar and the 120 ml (6 tablespoons) (9 tablespoons) of water into a
heavy-based pan. Allow the sugar to dissolve over a very low heat and when completely dissolved bring to a rapid boil.
Allow to cook without stirring until it is a

rich caramel colour. Remove from the heat, cover the hand holding the pan, and quickly add the extra water. Cool slightly before adding this thick syrup to the cooled custard. Continue with basic recipe.

Lemon ice cream
2 large lemons
50 gm (2 oz) (4 tablespoons) caster sugar

Peel the lemon rinds thinly and add to the milk for the custard and bring very slowly to the boil. Leave on one side to infuse for 10 minutes. Strain the milk and continue with the recipe to make the custard. Allow the custard to cool then stir in the strained juice of the lemons and the sugar. Continue with basic recipe.

Selling cream

It is not worth trying to sell cream unless you have a small dairy herd of cows and are equipped with a small dairy and centrifugal separator. You are also required to have a cream license from the department of agriculture who will want to inspect the premises and test the milk, cream and water supply. There is no commercial market for goats' milk cream and anyway, the process of separation is not economic in relation to the yield obtained.

There is a market for Jersey cream and anyone interested in this is advised to contact the local Dairy Husbandry Officer of the Department of Agriculture, as well as the Jersey Cattle Society.

The cream will need to be described on the packaging and the popular way of selling Jersey cream is in small pots with snap-on lids or foil tops. The packaging methods described for yoghurt on page 29 are similar, and the appropriate equipment is available from suppliers. The Jersey Cattle Society also sells appropriately illustrated cream containers and lids.

Note: Use either metric, *or* imperial *or* American measures in each recipe.

6

BUTTER

Butter is produced when cream is moved around rapidly until it reaches 'breaking' point. This is the point at which the fat droplets, which up to then have been kept separate from each other by the other components, coalesce or merge together. The best cream for buttermaking is that which has a butterfat content of 30%; a lighter cream will have a greater proportion of liquid which will act as a barrier between the fat droplets.

Before butter is made, the cream should be left to ripen. Ripening is brought about by bacteria acting on the lactose sugar and converting it into lactic acid. This is a natural process and will take place if the cream is left for a time. The optimum degree of acidity for buttermaking is 0.5–0.6%, and this can be determined by the use of a Lloyd's acidmeter (Fig. 16). Once you have made butter a few times, you will know from experience whether or not it is ready for churning.

Commercially, lactic culture starters are used to ensure that the acidity is correct, and these are added after pasteurization of the cream. The pasteurization ensures that any other bacteria are killed off. For home buttermaking, commercial starters are not necessary, nor is it necessary to pasteurize.

Buttermaking

Equipment

It is possible to make butter just by putting cream in a stoppered bottle and shaking it up. The drawback is that it takes longer, and that particular action is very tiring after a while. Some people use an electric mixer, but the difficulty here is in confining the cream in one place – if you use enough cream to make a decent portion of butter, it is more than most mixers can cope with, and cream flies all over the room when you switch on.

A small hand churn is adequate for most people's requirements and these are available from suppliers of dairying equipment. For larger scale production, electric models are available, including some which will churn whole milk. These are more applicable to those with large quantities of milk.

A dairy thermometer is a good investment, for it can be used not only for buttermaking, but for yoghurt and cheese production. It has the ability to float, which is particularly useful in cheesemaking, and the optimum temperatures for churning, cheesemaking and pasteurizing are marked on the scale.

Scotch hands or butter pats are useful for working the butter. This is the process of squeezing out any surplus liquid that remains after the butter has been made and removed from the churn. A palette knife will suffice as an alternative to the butter pats.

Butter moulds are extremely nice to give your butter a finish, particularly if you can get hold of one with a design for impressing on the top of the butter.

When I was a child in Wales, we made butter twice a week – once every three working days, but never on a Sunday! We

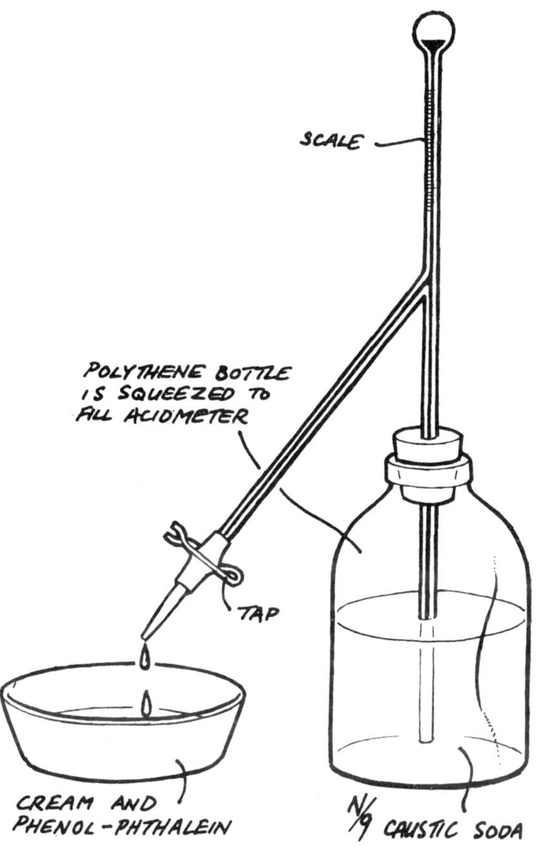

Fig. 16 *Method of using the Lloyd's acidmeter to determine acidity*

used an 'end-over-end' churn, which made a lovely swishing sound as the milk tumbled. My father produced the first mechanized churn in the Lleyn peninsula, by using his old motor-bike to power it. It must have been the only kick-started butter churn in existence.

Making butter is not only easy but thoroughly satisfying. I've made it from the milk of Welsh Black and Jersey cows and Anglo-Nubian goats. My ideal is Jersey cow milk with its rich supply of cream which readily separates onto the surface. The legal lower limit for fat in milk is 3%. The fat content varies considerably from breed to breed, and is also dependent on time of year, period of lactation, time of milking and regularity of milking.

Some people say that as goats' milk is naturally homogenized, you can't make butter unless you separate the cream with a mechanical separator. This is often true with goats' milk of a poor butterfat content, but with milk from Anglo-Nubians or other good milk strains the cream will rise to the surface if it is left. Heating the milk slightly

Fig. 17 *Buttermaking equipment*

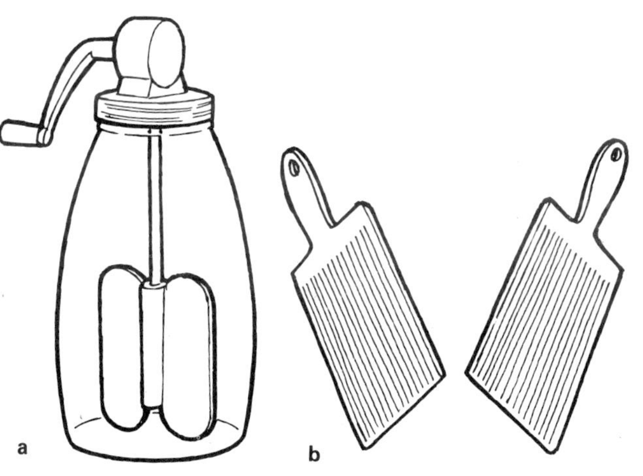

a   Blow butter churn
b   Scotch hands

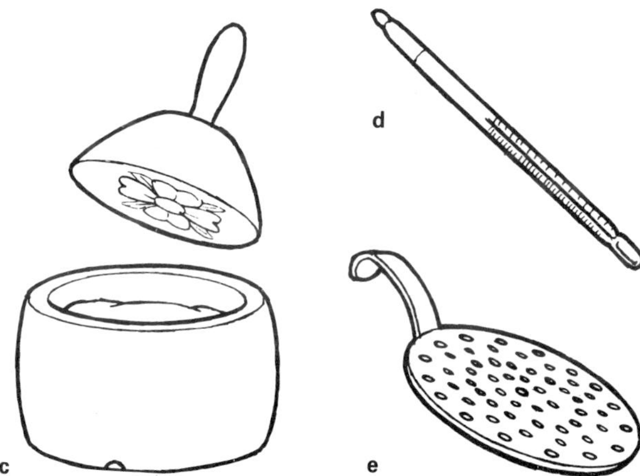

c   Wooden butter mould with drainage hole and butter print
d   Floating dairy thermometer
e   Hand skimmer

Fig. 18 *Step-by-step buttermaking*

causes more of the cream to rise, enabling it to be skimmed from the surface.

You should ideally have your own source of milk for buttermaking, but there is no reason why you shouldn't use milk from the milkman, provided it is not homogenized – except for the cost. (Homogenized milk is milk which has had the fat globules broken up into small particles which are distributed throughout the milk.) Butter can be made from the whole milk or from the separated cream but if you use whole milk you will need a large churn such as the 'end-over-end' to cater for the volume of liquid.

Churning temperature

The temperature needed is just below that of room temperature (or just above if the room is particularly cold) and not, as is commonly supposed, at a hard-and-fast degree. The Ministry of Agriculture gives the following as an indication of suitable churning temperatures:

Room Temperature	Temperature for Churning
18°C (65°F)	11°C (52°F)
14°C (56°F)	13°C (55°F)
10°C (50°F)	15°C (59°F)

I stand my cream on the edge of the stove to warm up a bit if necessary, but unless you have an experienced finger, this is where a dairy thermometer comes in useful.

Churning

A wide range of churning methods were used at one time, and in my home village of Tydweiliog in Wales, in my grandmother's day, both horse and dog power were utilized, as well as children and women. Nowadays, the choice of churn extends from a large glass jar with a screw-top lid, to a full size electric churn. The container should not be more than one third full if it is a full size churn or more than half full if it is a small one. Churning can take anything from fifteen minutes to an hour depending on the temperature, acidity, quantity and power source.

We have a hand churn and a small electric one and my family once had a race using them. My husband took the hand churn and the children used the electric one. It was virtually neck-and-neck, taking twenty minutes, with the electric churn only seconds ahead at the end, but my husband was red-faced and panting as a result.

A little water may be added to the cream to get the consistency right for churning. If the back of a wooden spoon can be just seen when cream is poured over it, then it's right – but take care not to add too much water. If the cream is a bit cold, a quick way of warming it up is to add a little hot water at this stage. Whatever churn you use, the process is the same. Nothing seems to be happening until suddenly, there they are – the grains of butter, and the liquid begins to look watery. This is called the 'breaking' stage.

Churning unseparated milk

This was the normal way of making butter on many farms. In addition to the usual 'end-over-end' churn there were sometimes older types still in use. One of my uncles, for example, was still using a traditional Welsh 'rocker' churn when I was a child. It was dark oak and squeaked as it rocked from side to side. One of my childish fears was that I would have an overwhelming impulse to put my finger under the rocker and get it caught, but I never did. A few years before she died, my mother dictated the following description of how she made butter from unseparated milk. I have translated it from the Welsh.

Years ago, in North-West Wales, we made *menyn bach*, salted butter made from whole milk, churned in an end-over-end churn. After taking enough milk for the day's use, the remainder of the milking was set aside. It was about two containersful twice a week. As the churn held approximately two containersful when two-thirds full, churning took place twice a week. The sour milk was poured into an end-over-end churn and the lid clamped down. During the first five minutes it was necessary to stop and release air from the valve in the lid about three times, then churning continued for about twenty minutes. After this time, grains the size of coarse sand appeared, sliding on the window of the churn lid. At this stage, churning ceased and the churn was rocked gently from side to side so that the particles began to clump together in the centre of the liquid.

The churn was opened and the butter gathered up with a *cwpan dena* a thin shallow wooden saucer, and put into a *nou* which was a round wooden basin. The wooden saucer was used to squeeze the liquid from the butter which was then beaten hard in the bowl, and rinsed with cold water to clear away the butter milk – *llaeth enwyn*. Salt was added and the butter beaten hard again with the saucer. Finally it was formed into a round, put on greaseproof paper and imprinted on the top with a wooden print block. The yield was about 3 lbs of *menyn bach* – the most delicious butter I have ever tasted, which, with new potatoes and a glass of buttermilk made a feast worthy of a king, and the surplus buttermilk grew fine healthy pigs.

It is possible to buy electric powered churns which will churn whole milk, although most people now choose a smaller model for the churning of cream. Hand-operated and electric models are available. Beware of buying second-hand wooden churns at farm auctions. If they have been used for mixing farm chemicals (as some of them were after farm buttermaking ceased) the wood may be impregnated with toxins. Many of these old churns are not worth buying anyway because antique dealers have pushed the prices up to an unrealistic level. It is much better to buy a new hand model from a dairy supplier and leave the nostalgia seekers to enjoy their useless and expensive memorabilia.

Top of the milk butter

The following is a description of how an urban dweller, Mr Bob Rogers of Swansea in Wales, makes butter from the cream at the top of bottled milk.

We have two pints of silver top milk a day and make about 170 gm (6 oz) of butter a week. We open yesterday's milk, pour off the cream top into a basin and each time a bottle of milk is opened we add to it. It is kept in the refrigerator until ready for churning. We use a glass Blow churn and, from start to finish, churning takes about ten minutes. The buttermilk is strained into a jug. It is lovely to drink or for use in cooking. During the straining the butter grains are held back with a table fork. Cold water is added, churned and drained twice. We then tip the butter onto a board and work with a knife, pouring on water to clear and adding salt to taste.

Washing

The buttermilk is strained off through muslin and cold water is added about 2° colder than the churning temperature. Do not throw the buttermilk away. It is a fine drink, can be made into buttermilk cheese, added to mashed potato, used for making scones or fed to your animals. The churn is turned again and the process is repeated until all the buttermilk has been washed away, and the water remains clear.

Salting

Being Welsh, I like my butter highly salted, but a French friend tells me that I am a barbarian. If you are a barbarian too then start with a teaspoon of salt sifted over 450 gms (1 lb) (2 cups) of butter, then add more to taste. If you add too much you can always wash it out again. Salt adds to the keeping qualities of the butter, but if you use it in cooking, in a recipe that does not require salt, then you may spoil your cake.

Working

The process of 'working' is to get rid of surplus moisture. Butter intended for sale must not contain more than 16% water. For working you need a wooden board such as a bread board, some Scotch hands (see fig. 17) and a piece of muslin cloth or clean linen. The butter is pressed flat on the board until drops of moisture appear. These are wiped off with the muslin, and the butter is then rolled up into a Swiss-roll and pressed again. Repeat, making the Swiss-roll from the other sides. This is repeated until no drops of moisture are seen. Scotch hands are available from dairy equipment suppliers but a cake spatula or knife will suffice, although more difficult to use. The aim is to work in such a way that the 'grain' of the butter is not lost by rubbing. You can, of course, use your hands but it's more difficult to retain the 'grain'.

It is worth mentioning that working the butter is straightforward on a small scale, but if you want to make more than a small quantity at a time then difficulties arise at this stage. The problem is having a sufficient working area and being able to apply enough pressure over a big enough surface area to remove the water. Traditionally, a butter worker was used on the farm. This was usually made of wood and had a grooved roller so that, as the handle was turned, the butter was pushed through the worker, while having a considerable amount of surface area exposed. Unfortunately, these are no longer manufactured and the only commercial equipment available is on far too big a scale to be relevant to the domestic scale. Commercially, the working process is brought about through the use of augers which press the butter through perforated plates before being extruded through nozzles into a packaging machine.

Shaping

Once the surplus moisture is removed, the butter can be shaped into a pat, and if you are lucky enough to have a pattern mould the design can be impressed on the top. I use a wooden bowl with a hole at the bottom and a mould with a handle which presses its pattern down on the butter. Any remaining water is then squeezed out through the hole at the bottom of the bowl. Finally, the butter can either be put into a dish or wrapped in greaseproof paper with cellophane on the top.

Colouring

Sometimes butter is pale, particularly if the cow is late in lactation. Certain breeds, like Friesians produce paler butter anyway because the fat content is lower than, for example, Guernseys or Jerseys. Goats' milk butter is pure white and looks rather like lard, although the taste is the same as any other butter. If you want to make your butter yellower you can add annatto either at the 'breaking' stage or during 'working', although for domestic use it is hardly necessary. Annatto is a harmless colouring substance obtained from the seeds of *Bixa Orellana*, which grows wild in the West Indies and South America. The colouring matter is dissolved in refined vegetable oil which transmits the colour to butter without colouring the buttermilk. It is available from most dairy suppliers.

Storing butter

Butter will keep for months in a deep freezer, but beware of not removing surplus moisture. 28 gm (1 oz) (¼ cup) of salt to 450 gm (1 lb) (2 cups) of butter will keep it from going rancid, particularly if it is stored in earthenware pots in a cool place.

Buttermaking problems

Butter doesn't come?
Check the temperature, and if necessary warm it up a bit.

Butter has a fishy taste?
Did you 'work' the butter on the board where you'd been filleting fish? If you make butter in an old dairy and all the butter you make tastes fishy, it could be caused by a small mould called *Oidium lactis* which lives in the woodwork of old dairies. Washing the woodwork with lime eradicates it, but as it is rare these days you are not likely to come across it.

Butter has taints?
Make sure that the cream is left to settle in a cool place, covered up, and well away from such things as onions. Ensure that the dairy animal is not eating plants which

taint milk such as too many turnips or garlic. Make sure that all utensils are clean and sterile. Boiling water should be used for all the equipment.

Butter has cheesey flavour?
The most common cause is having conditions which are too acidic. Ensure that all equipment is sterile and next time, do not leave the cream to ripen as long.

Rancid butter?
Rancidity is caused by hydrolysis of the fat into fatty acids and glycerine and has a distinctive aroma. Dirty equipment or insufficient washing of the butter could be the culprit. Disregard the irresponsible advice which is given in some books about how it is possible to sweeten rancid butter by various methods. The only place for it, if you wish to steer clear of possible food poisoning, is in the bin.

Butter recipes

Sweet Butter
This is butter that is made from freshly separated cream which has not been allowed to 'ripen' at all. It takes longer to churn.

Herb Butter
Finely chopped fresh herbs or powdered dried herbs are mixed in with the butter. I prefer to use fresh herbs, and particularly good are chives, parsley, mint, lemon balm, thyme, sage. The quantity used varies with each herb, depending on its strength – less sage than parsley for example, but it is a matter of taste, so experiment. It needs to be left for a few hours before using: this allows time for the herb to permeate the butter.

Brandy Butter
450 gm (1 lb) (2 cups) unsalted butter
700 gm (1½ lb) (5 cups) sieved icing sugar,
240 ml (12 tablespoons) (18 tablespoons) brandy.

Cream the butter and gradually work in the icing sugar and brandy. Leave in a cool place to harden before use. It is delicious with Christmas pudding. (Rum may be used if preferred.)

Butter icing
This is nice as a filling for sponge cakes or to go on top of a special occasion cake. Beat 50 g (2 oz) (4 tablespoons) of butter in a bowl until light, then sift 110 gm (4 oz) of icing sugar on it, working it in with a fork. A few drops of vanilla can be added or any of the following flavour: 1 dessertspoon cocoa; 1 dessertspoonful coffee essence.

7

CHEESE

Cheesemaking is an ancient art. No-one is certain where it originated, but there is evidence to suggest that it was made in Macedonia, the Egypt of the Pharoahs and in many parts of Asia before it ever appeared in Europe. Wherever it came from there would be few who would disagree with the ancient Greeks that it is 'a gift from the gods'.

The process of cheesemaking involves modifying the milk protein casein, usually through the use of the coagulant rennet, so that the solids separate from the liquid – the curds and whey. The curds are subjected to a number of different processes such as acidifying, heating, pressing and so on, and ultimately, cheese is produced. There are, however, an infinite number of variables within this broad framework, which are related to type of milk, creaminess, degree of acidity, type of bacteria and moulds, length of ripening and many other factors. For example, sheep's milk is used to make Roquefort, but if cow's milk is used, the resulting cheese will be Bleu des Causses – even though production methods have been identical. The cheesemaker has to determine the best conditions for whatever cheese he or she is making. He can control the amount of cream by skimming some off, or alternatively, by adding some more. He can control the acidity by ripening the milk for a certain length of time, or he can utilize a starter – a culture of lactic acid-producing bacteria. The amount of rennet used determines the degree of ripening, as well as being the coagulating factor. The ripening time has to be determined, and this can vary from a few months to years. The temperature and degree of acidity when the rennet is added also determines whether the cheese will be soft or hard. A slight variation in any of these factors will produce a different cheese. So, the field is wide open for experiment.

Types of cheese

There appears to be no universally agreed system of classifying cheeses. Some authorities categorize them according to the ripening time, whether they are pressed or not, or even by nationality. As this book is primarily for the home cheesemaker, I have classified them according to their production methods.

Soft, naturally soured cheese

These are cheeses which separate as a result of natural souring where the increasing acidity acts on the milk protein casein. An example of one of these would be a pot cheese where after curdling the liquid whey is poured away and the curds left to drain in a suspended cloth.

Soft renneted cheeses

These are cheeses which have rennet added to them before the natural acidity is very high, so that curdling takes place more quickly than would otherwise have occurred. An example is Coulommier cheese.

Soft ripened cheeses

A soft cheese may be eaten fresh or, if it has been subjected to a partial preservation technique, it may be left for several weeks to ripen. An example is a Brie-type cheese which is slightly salted before being ripened for about three weeks.

Semi-hard cheeses

As the name implies these are cheeses which have been subjected to a certain amount of pressure, but not enough to destroy the light, crumbly texture of the final product. Caerphilly is an example of such a cheese.

Hard, pressed cheeses

These are cheeses such as Cheddar which have been subjected to considerable pressure in order to remove a large proportion of liquid, before they are left to ripen in storage.

Innoculated cheese

These cheeses have had innoculants such as moulds introduced into them (usually by means of stainless steel needles) so that a particular flavour and aroma is produced. Examples are Blue Stilton, Danish Blue and Lymeswold.

The elements of cheesemaking

It has already been said that cheese curds are produced when lactic acid makes the milk protein casein coagulate, but as this process has to be controlled depending on the type of cheese being produced, it is worth looking at what is involved in greater detail. This involves rennet, testing acidity and the use of culture starters.

Rennet

Rennet is a coagulating or curdling agent obtained from the stomach of a calf. It contains the enzymes rennin and pepsin in close combination, and when added to milk, acts on the milk protein casein, bringing about a separation into curds and whey – the solids and the liquid. If raw milk is left for a few days it will curdle anyway. This is because bacteria act on the milk sugar lactose, forming lactic acid. The longer it is left, the more acid it becomes, until it finally curdles. In cheesemaking however, it is not always convenient to have coagulation at a very acid level; some cheeses require coagulation while the milk is sweet. This is where rennet comes it. It is easily available at dairy suppliers, and will store well provided it is kept in a cool place and out of the light. It is in fact the light more than anything which makes it lose its effect, and reputable suppliers will always sell it in darkened bottles. For soft cheeses, the weak rennet sold in chemists will suffice, but for hard or semi-hard cheeses you do need the strong cheese rennet. The more rennet is added, the firmer will be the cheese. Some vegetarians or Orthodox Jews may object to the use of calf's rennet, and special vegetarian rennet is available commercially. The pressed juice of certain plants has also been used in the past. Sometimes the wild flower, Ladies' Bedstraw, is still called Cheese Flower as an indication of this. For soft cheeses, lemon juice or even a small amount of vinegar will coagulate the curd, but these do impart a strong flavour.

Testing the acidity

At different stages of cheesemaking, it is necessary to know the level of acidity of milk, cream or whey. For example, when making a Cheshire cheese the acidity or level of lactic acid in the milk should be 0.20% before rennet is added. Perfectly good cheeses can be made without as detailed a technique as this, but if a cheese is to be duplicated exactly a second time, it will need to be made in precisely the same way. What a determination of the acidity does is to help you provide the optimum conditions. If these are recorded as you go, you will be able to produce the same cheese next time.

The Lloyd's acidmeter is the piece of

apparatus for determining the degree of acidity of milk, cream or whey. The following is an explanation of how to test the acidity of cream (in grams) but the technique is exactly the same for milk or whey.

Ten grams of cream are put in a white dish and 2 or 3 drops of phenolphthalein solution are added. Caustic soda of N/9 strength is added, drop by drop, by opening the clip of the burette, until a very faint pink tinge appears in the cream phenolphthalein mixture. Each millilitre (ml) of the caustic soda will neutralize .01 of a gram of lactic acid. If this amount were contained in 10 grams of the cream the percentage would be 0.1. If when reading the results, 5 of the large divisions and 6 small ones have been used to produce the pink colour, then the cream contains 0.56% of lactic acid. (An N/9 solution of caustic soda is made by dissolving 4.5 grams of pure caustic soda in one litre of distilled water.)

Starters
There is a lot of confusion about 'starters'. A 'starter' is a culture of lactic-acid-producing bacteria which provides the milk with an optimum level of acidity for cheesemaking. It also gives it flavour and aroma, as well as a firmer curd. It is possible to obtain a commercial starter in liquid form or in a freeze-dried sachet. It is also possible to make your own in several ways.

Home-made starters One way of making your own starter is to take a pint of fresh, filtered but unpasteurized milk from your own dairy animal. Place it in a sterilized bowl, cover and leave in a warm place for two to three days. By then it should have curdled fairly firmly, with a clean acid smell. If there are signs of mould, flakiness or a smell of yeast or cheese, then discard it. It is essential to have good quality milk and hygienic conditions. Do not try to produce a starter with previously pasteurized milk. This includes bottled milk from a dairy because this will have been pasteurized.

Fresh buttermilk left over from butter-making can be used as a starter, if it is ripened in similar conditions to those detailed above. Cultured buttermilk bought at a shop may be used, but it should be used straight away.

It is also possible to make your own from a small piece of shop-bought cheese or a piece of your own previously made one. The technique is simple and is based on the fact that a specific cheese will provide a starter suitable for a similar type of cheese. In other words, if you want to make a Cheddar-type cheese, use a piece of existing Cheddar for your starter. Similarly, if you want to have a go at making a Brie-type cheese, then use a small piece of shop-bought one. (Do not try to sell your cheeses under these specific names, however, or you may be contravening the Trades Description Act.) The piece of cheese used should always come from the inside, just under the rind.

Ensure that all utensils are scalded with boiling water to kill off bacteria before you begin and follow the steps described in figure 19.

One point to remember is that if you are using shop-bought cheeses for your starters, avoid ones which are processed. You need to use ones which are still maturing and so have live bacteria still working. When adding water to the cheese, make sure that it is not too hot otherwise you may kill off the bacteria. It should be no higher than 32°C (90°F) although it does not matter too much if it goes below this. Do not try to store your starter for use on a future occasion. Use it straight away. It is easy enough to make a fresh one in this way whenever you need it. Do make sure that after making your cheese you leave it to ripen properly so that the full aroma and flavour develop.

Commercial starters These are available in liquid form or as freeze-dried sachets. The freeze-dried form will store in the deep freezer for six months. Whichever form the

1 Take a teaspoonful of cheese from just below the
   rind.

2 Pour on about half a cupful of water which has been
   boiled and then cooled to no higher than 32°C (90°F).
   Stir up the cheesy mixture, breaking it up as much
   as possible with a fork.

3 Leave to steep for about half an hour in a warm
   place such as an airing cupboard. Keep covered,
   either with a piece of muslin or a saucer so that the
   dust does not get in.

Fig. 19 *Making your own cheese starter*

4 Strain through muslin or a tea strainer so that any
   remaining solid lumps can be discarded. The
   resulting liquid is now ready to be used as a starter
   for your cheese.

starter is in, it will need to be propagated for use on a continuing basis. There are three stages involved in this.

1 Preparing sterilized milk

Using either boilable polythene bottles with screw caps or tonic-water bottles, fill up to the neck with milk that has previously been boiled and cooled slightly. Screw on the tops, then unscrew by half a turn. Put the bottles on a rack, trivet or crumpled cloth in a deep pan filled with water to the level of the milk. Put the lid on the pan and bring to the boil. Simmer for 1–2 hours, then allow to cool naturally. Tighten the screw caps and place in the refrigerator.

2 Innoculating the starter culture

You are now ready to innoculate a prepared bottle of sterilized milk with a live culture from a bought starter. In order to displace the bacteria-laden air immediately where you are working, it is a good idea to either have naked flame like a gas burner on a gas stove, or a saucepan of boiling water with steam rising from it. (Take care not to burn or scald yourself.) Shake the bought starter and loosen the caps on both bottles. Take off the cap of the shop starter, doing so in the displaced air, and pour a teaspoonful into a sterilized spoon. Immediately transfer this into the bottle of sterilized milk, and quickly seal. The original bottle of starter culture can be used for cheesemaking straight away, but the newly innoculated one will need to be incubated. If, however, it is not going to be used for some while, it can be put in the deep freezer at this stage and incubated immediately before use. Polythene bottles are essential for freezing for glass ones will crack.

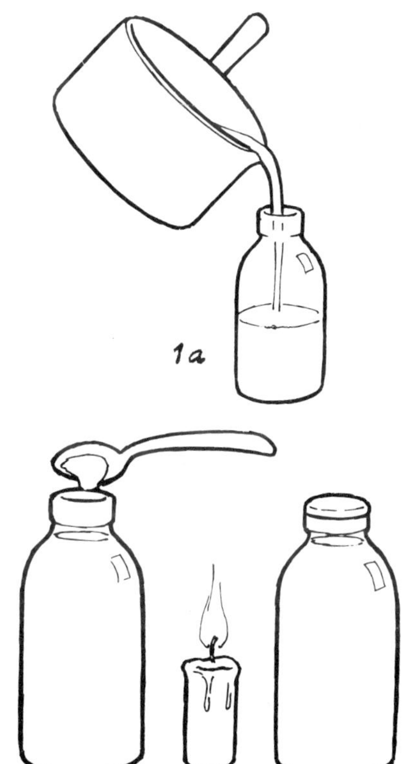

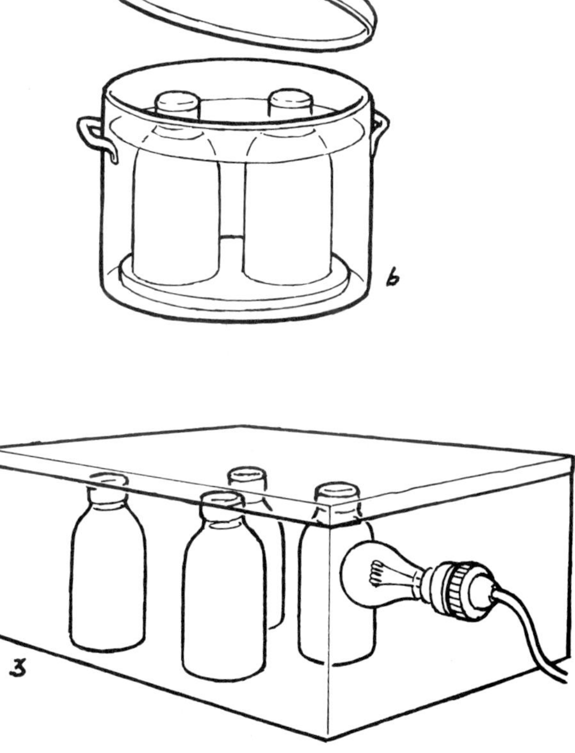

Fig. 20 *Propagation of starter*
1a *and* **b** *Preparing the sterilized milk.*
2 *Inoculating the starter culture.*
3 *Incubating the starter culture.*

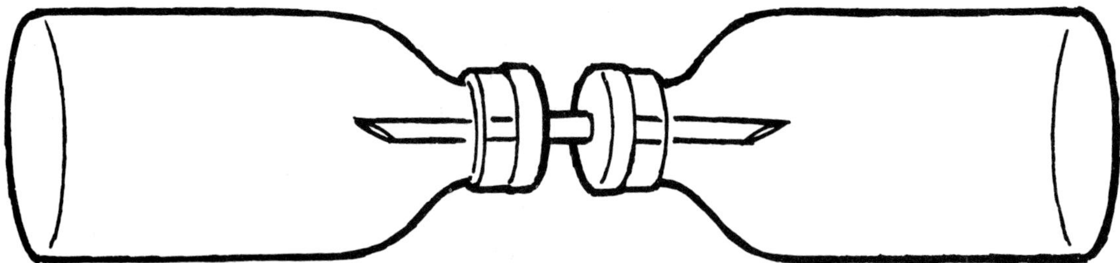

Fig. 21 *Lewis bottles*

3 Incubating the starter culture

Keep the new bottles at a temperature of 21°C (70°F) for 8–12 hours or until set. An airing cupboard is usually ideal for this. Other suitable places are egg incubators and plant propagators provided the temperature can be adjusted low enough. As soon as the milk has set it can be put in the refrigerator until used. A home-made incubator is easily made, and the size of the bulb used depends on the size of the box. In order to find this out you will need to experiment beforehand with a thermometer in the box. (You can have different bulbs for different purposes, so that a high watt bulb gives you an egg incubator and a lower one gives you a starter incubator.)

There is always the possibility of unwanted bacteria getting into your starter during the innoculation stage, and a method of lessening the risk is to use Lewis bottles (Fig. 21). These are polythene bottles with special rubber seals. The culture is transferred from one bottle to another with a hollow, double-ended innoculating needle, and the bottles themselves remain airtight. Although the rubber seals are punctured by the needle, they reseal afterwards and can be used many times.

The freeze-dried starters mentioned earlier are the ones normally supplied to large-scale cheese manufacturers, and the accompanying instructions suggest adding the contents of the sachet to 1–2 litres (2–4 pints) of milk. As this amount would be too much for the home cheesemaker to use at any one time, it is a good idea to make up a larger number of small polythene bottles for storage in the deep freezer.

Lactic acid is also available from chemists and can be used but it does not impart as good an 'aroma' to the cheese as a 'starter'.

Advice to first time cheesemakers

1 Unless you have your own source of milk, do not expect to be able to save money by making your own hard cheese. It is cheaper to buy it in the shop. Nevertheless the interest and enjoyment of making cheese will be considerable and this is an important point. Soft cheese is another matter. You will almost certainly save money by making your own soft cheeses, particularly if they are the more exotic ones which are expensive to buy.

2 Do not try to make a hard cheese without having first experimented with soft cheeses, and 'working your way up'.

3 You will have more consistent results by using a commercial 'starter' – a culture of the correct lactic acid-producing bacteria. After gaining some experience with these you can always experiment with making your own starters.

4 The weak rennet sold by chemists for junket is all right for soft cheeses but for hard cheeses it is essential to have the proper cheese-making rennet.

5 Do not try to guess temperatures, for they can be crucial at different stages of cheesemaking. Use a proper dairying thermometer.

6 Do not use pieces of builder's pipe as cheese moulds (suggested in some books).

The resins from that type of plastic are toxic.

7 Do let your hard cheeses ripen for the recommended times before eating them. Something that tastes like polystyrene can be transformed by a few month's ripening.

8 Do buy a proper cheese press if you are going to be making quite a few hard cheeses. They are not that expensive and will produce much better results.

Cheesemaking equipment

As the interest in home cheesemaking has increased, so equipment has become more easily available, and some items which have been unavailable for many years are now either being manufactured again or being imported. The amount of equipment you need really depends on the scale of your activities but if you want to go into hard cheesemaking as well as soft, the following are pretty important.

Double saucepan Unless you can heat your milk in a container in a waterbath, it is very difficult to control the temperature changes. The water around the inner pan ensures that any temperature changes are gradual. These are easily available at kitchen equipment or dairy suppliers. They are usually aluminium, although stainless steel ones are available. These are more expensive but good value for they last a lifetime.

Floating dairy thermometer This really is essential. You need to be able to notice even slight variations in temperature, and unless you have a floating thermometer for use in making something like a large Cheddar cheese, it could mean your standing holding it in the milk for half an hour at a time. In the USA, there are useful clip-on dairy thermometers which can be attached to the side of the container, but I have not seen these on sale anywhere else.

Rennet Rennet which contains the enzymes rennin and pepsin, acts as a coagulating or curdling agent for setting the milk (see page 48).

Lactic ferment culture starter This is a culture of the appropriate bacteria which will ensure that your milk is at the optimum level of acidity (see page 49).

Knives for cutting curd Curd knives are available from suppliers but a long palette knife will do.

Cheese cloth This is a close-textured cloth for draining curds. It ensures that the fat particles are not lost.

Butter muslin This is used for lining cheesemoulds before the curd is put in for pressing, and also for bandaging pressed cheeses.

Cheese moulds After many years, cheese moulds are available again, and the range of different types is continually increasing, as new ones are either manufactured or imported. At one time, most cheeses had their own distinctive moulds, but many moulds can be used for a range of different cheeses for example, the Colwick cheese mould imported from Switzerland can be used for making many soft cheeses that do not require pressing. The best ones are made from stainless steel or high density plastic. You can make your own from plastic food containers but avoid metal.

Cheese mats and trays Cane mats are useful for soft cheesemaking but the cane place mats which look very similar should on no account be used. Several people have tried this and found that their cheeses become discoloured, and the mats themselves disintegrate when sterilized. Plastic mats are also available. Cheesetrays are useful as draining trays for some of the smaller moulds, and will keep the whey from running everywhere.

Fig. 22 *Cheesemaking equipment*

a Double saucepan
b Floating dairy thermometer
c Lactic ferment culture starter
d Rennet
e Curd-cutting knives
f Muslin
g Cheese mat
h Cheese tray
i Measuring jug
j Ladle
k Spoon
l Cheese press
m Cheese moulds

Cheese press This really is a must for hard cheesemaking and there are several suitable ones on the market. A particularly good one is the one manufactured with its own stainless steel moulds, drip tray and beechwood follower.

Miscellaneous Also necessary will be miscellaneous items such as measuring jugs, ladles, wooden spoons.

The principles of hard cheese-making

Different hard cheeses will vary slightly in the method of manufacture but allowing for these variations (which are detailed in specific recipes in the last chapter), the basic steps are as follows:

1 *Heat treatment* This is where milk is heated to 68°C (155°F) then cooled to the optimum temperature for cheesemaking of 30°C (86°F). The heating destroys unwanted bacteria.

2 *Adding the starter* The appropriate amount of starter (commercial or home-made) is added so that the desirable bacteria will produce lactic acid at the optimum temperature of 30°C (86°F). The milk is then left for a given time, until the required acidity is reached.

3 *Renneting* The appropriate amount of cheese rennet is added. The usual practice is to dilute it in a given volume of previously boiled and cooled water before stirring it in. Top stirring of the milk may be necessary to prevent all the cream collecting at the top, until coagulation begins, but this depends on the particular recipe.

4 *Setting* The curd is normally ready when it is firm to the touch and does not leave a milk stain on the back of the finger. The amount of time left before cutting depends on the required acidity for a particular cheese.

5 *Cutting the curd* This is where the curd is cut first into 1 cm (½ in) strips, then at right angles to form 1 cm (½ in) square columns. Unless a curd knife is available to make horizontal cuts, a palette knife can be used to make diagonal cuts until individual squares of curd are produced. The curd is then loosened around the walls of the pan and left until whey appears at the top. Again the time will vary depending on the relative acidity of the whey required at this stage.

6 *Scalding* This is where the temperature of the curds and whey is raised slowly while stirring of the curds takes place by hand. The usual temperature increase is 38°C (100°F) achieved over a period of half an hour.

7 *Pitching* This is the process of giving the whey a quick circular stir so that it whirls round while at the same time allowing the curds to sink to the bottom and collect at a central point. The heat is turned off at this point and the pan left for the appropriate time. This is usually about thirty minutes.

8 *Running the whey* As much of the liquid whey as possible is ladled out, then a sterilized cloth is placed over a stainless steel bucket or large basin and the curds are tipped in. The cloth is then made into a bundle by tying a Stilton knot, where one corner is wound around the other three. The bundle is hung up or placed on a tray which is tilted at an angle to let the whey drain.

9 *Stacking or cheddaring* After a short period the bundle is untied when the curds have formed a mass. Cut this into four slices and place one on top of the other then cover with the cloth. After about fifteen minutes place the outer slices of the curd on the inside of the stack, and vice versa. Repeat this process several times until the curd resembles cooked breast of chicken when it is broken.

10 *Milling* This is the process of cutting the curd into small pieces. Traditionally a curd mill was used for this, but it is relatively easy to do it by hand. Different

Fig. 23 *Stages in making a hard cheese*

13 Tip curds into sterilized cloth

14 Leave bundle to drain

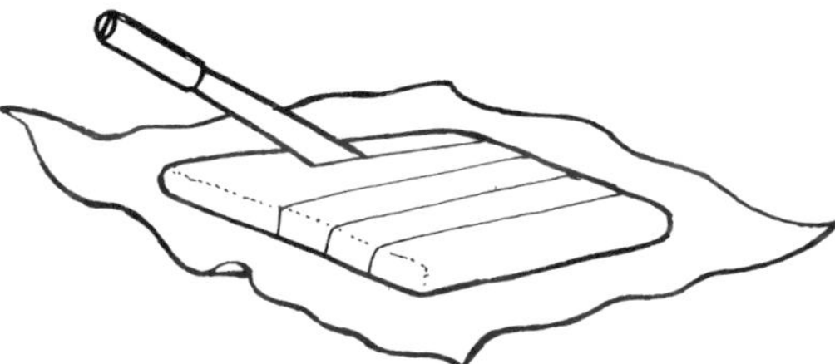

15 Cut curd into long slices

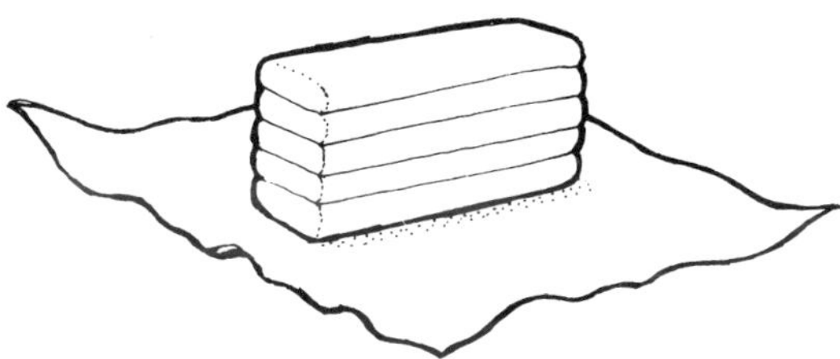

16 Stack slices of curd

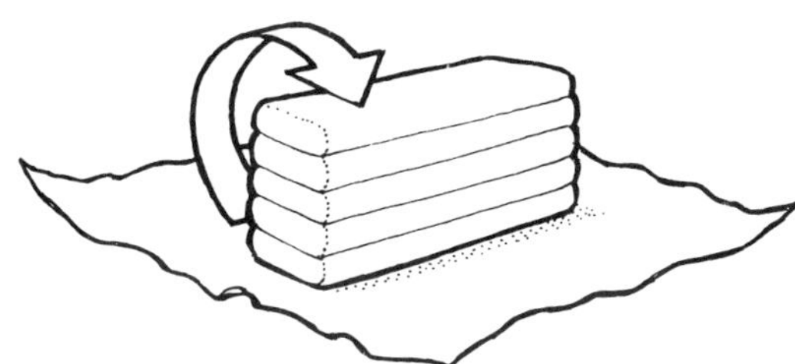

17 Restack curd slices several times

18 Break curd into pieces and add salt

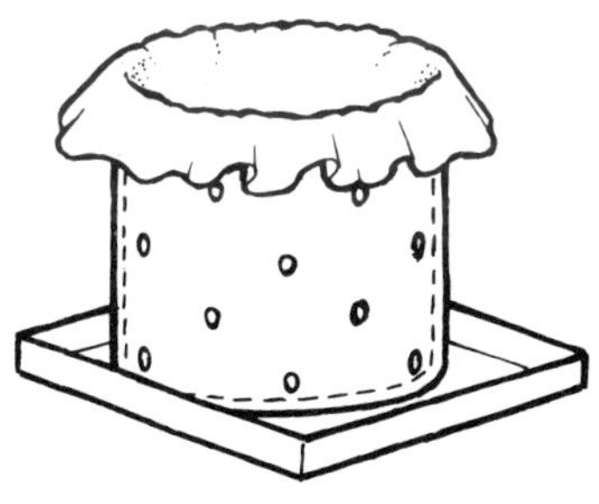

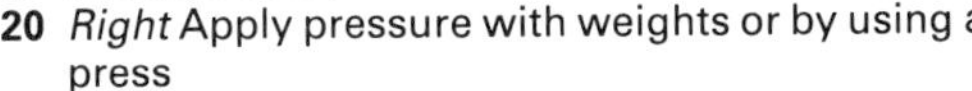

19 Put curd in moulds lined with muslin
20 *Right* Apply pressure with weights or by using a
press

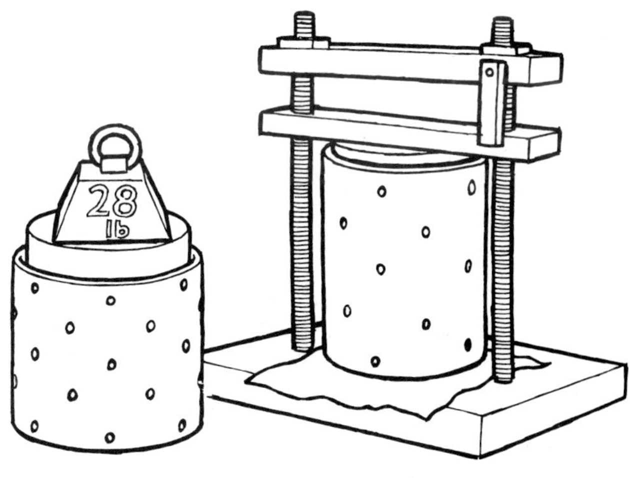

21   After pressing leave to cool and dry

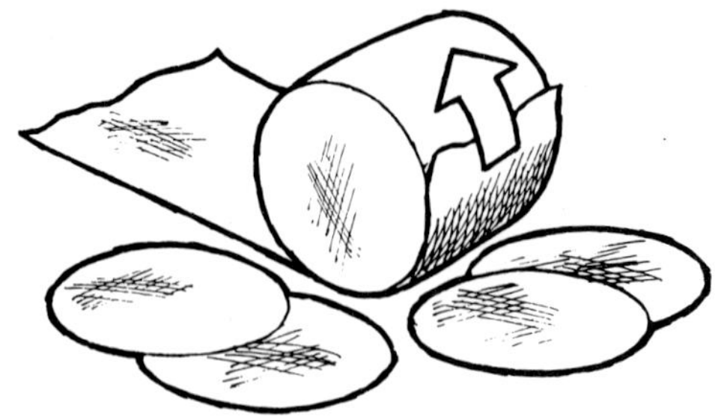

22   Bandage with muslin and cut end caps

23   Store between 10–16°C (50–60°F) on shelf or
suspended in a muslin bag

recipes call for different sized pieces but
generally they will be the size of a nutmeg.
The pieces are best placed on a tray so that
they are ready for the next stage.

11 *Salting*   At this stage salt is added to
the milled curds and the old way of doing it
was for two people to toss the curds while
holding the cloth at each end. On a small
scale it is easy enough to sprinkle the salt
onto the pieces, rolling them gently without
breaking them further. Again, the amount
of salt will depend on the specific recipe,
but as a general rule it is 25g (1 oz) salt to
1.3 kg (3 lb) curd.

12 *Moulding*   This is the process of lining
a mould with boiled muslin and ladling in
the curd until the mould is full. The corner
of the muslin is then folded over the top of
the cheese and it is ready for pressing.

13 *Pressing*   Once in the mould the curds
have a wooden 'follower' placed on top so
that when the mould is put in the press
there is a surface on which to exert an even
pressure. Pressing should take place gradually so that the whey is discarded without a substantial loss in the fat content of
the curds. Depending on the recipe, the
cheese will be taken out of the press and
turned several times so that pressure is
applied evenly.

It is possible to make your own press or
to use heavy weights balanced on a board,
but it is difficult to control the pressure
properly in this way and I would recommend buying a proper one such as the
Wheeler press which is widely available in
Britain, the USA and to a lesser extent in
Australia. It has stainless steel moulds and
drip tray while the wooden parts are made
of hard beech which is easy to clean.

14 *Cooling*   After the required period of
pressing the cheese is taken out of the press
and allowed to cool and dry for between
two to five days so that a rind begins to
form. It is uncovered at this stage and care
should be taken to ensure that no dust or
dirt affects it. It is also vulnerable to pests
such as houseflies and cheese mites. The
traditional way of affording protection was
to place the cheese on cane mats on a slate
shelf with a fine mesh cover rather like an
inverted umbrella over it. This allowed air

to enter and circulate freely. It is not difficult to make such a canopy from a wooden box frame with muslin stretched and tacked over it.

15 *Sealing* Once the rind has formed the cheese is sealed to prevent it becoming unduly desiccated while it is ripening. There are three ways of doing this:

Bandaging A muslin bandage not only protects the cheese but also holds it together in the event of 'blowing up'. This is when air holes which have not been dispersed during pressing expand and blow up the cheese until it resembles a football. This is another reason why cheeses are stored in relatively cool conditions so that the heat does not lead to expansion. The procedure for bandaging is as follows: cut a piece of muslin as wide as the depth of the cheese and one and a half times its circumference in length. Cut four circular pieces to act as caps for the top and bottom, but make them larger than the cheese so that they will fold over onto the sides. Using lard or flour paste stick two caps at each end then wrap the bandage firmly around the cheese, sticking it down as you proceed.

Waxing An alternative method is to wax the cheese (see Fig. 24). This is normally the custom with some of the semi-hard cheeses such as Edam or Gouda but it can be done with most semi-hard or hard cheeses made at home. Cheese wax is available from dairy suppliers and is pliable and easy to apply. Paraffin wax such as that used in candlemaking can be used but it is more rigid and therefore cracks more easily; the dairy wax is preferable. Cut the wax into small pieces and place in a bowl over hot water (just as you would if you were melting chocolate). Stir the wax every now and then to ensure that it is melting evenly. Dip the cheese into the liquid wax and coat thoroughly. It sets quickly so that the area where your fingers were touching can also be coated by rotating the cheese. If preferred, you can paint on the wax with a paint brush but this will probably need two coats.

Oiling A cheese can also be oiled with vegetable oil to provide a protective and

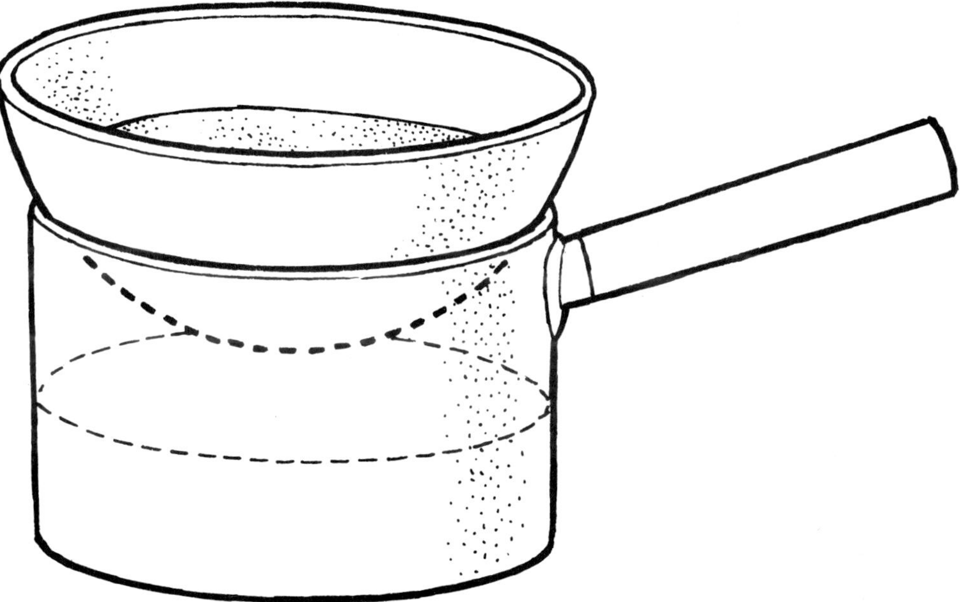

Fig. 24 *Melting wax for coating cheese*

anti-desiccating layer. It is not as effective as bandaging or waxing but is quick and perfectly satisfactory for cheeses which do not need to remain in storage very long.

16 *Ripening*   The last stage is often the all-important one. A cheese which is tasteless and bland when freshly made is full of flavour and body after its proper ripening period. Many shop cheeses are sold before they are properly ripened because of the high costs involved in an extensive ripening stage. Even Stilton is often sold after a two month storage time where traditionally it was never eaten until it was at least four months old. The correct storage temperature is between 10–16°C (50–60°F), although some cheeses may require different conditions. These are mentioned in the specific recipes. Wooden shelves are suitable for storing the cheeses which should be turned at frequent intervals. An alternative method of storage is to hang the cheese in a muslin bag so that the need for turning is dispensed with.

Growing moulds

There are two types of mould which are allowed to grow in ripening cheeses. These are the blue moulds found inside such cheeses as Roquefort, Danish Blue, Stilton and Lymeswold, and the white ones which grow outside cheeses such as Brie and Camembert.

Blue moulds   These are produced by the growth of *Penicillium roqueforti* which is available in the form of freeze-dried sachets of powdered spores. Application is either to the milk before the renneting stage or to the curds at the salting stage. For the home cheesemaker perfectly good results can be obtained by using a piece of shop-bought blue cheese, breaking it up into small particles and mixing with water as a starter (Fig. 19). Alternatively, small particles of the mould from a shop cheese can be sprinkled onto the curds at the salting stage. Once the cheese is shaped, the mould must have air in order to grow properly. The easiest way of ensuring this, is to make holes in the cheese with a sterilized stainless steel needle; a kebab skewer easily available in most kitchen suppliers is ideal for this, and is far preferable to the knitting needle advocated by some people. The earlier the air holes are made the better, and there is no need to wait until the cheese is out of the mould because the needle can be passed through the holes at the side of the mould, straight into the cheese. If the commercial spores have not been used and you have not used mould from a shop cheese by this stage, it is still not too late. Try dipping the needle in the mould from a shop cheese before making the air holes, so that the needle is introducing the mould at the same time as aerating the cheese.

Once the cheese is out of the stainless steel or plastic mould, pierce it from side to side and from top to bottom with holes about 3 cm (1 in) apart. During this time it should be stored at approximately 10°C (50°F) and will need frequent turning. A fairly high level of humidity is required and a convenient way of ensuring this is to place a large bowl of water close by. Where only one or two cheeses are involved the humidity can be assured by putting a plastic box or some other support in a bowl of water, with a board and cheese mat placed on top. The cheeses are above the water but have the benefit of the humidity. The blueing process should be apparent after two weeks. Any white or red moulds should be scraped off the outside. The cheese is usually ready to eat once the blue mould is apparent, but can be left longer if a more mature flavour is required. A light film of vegetable oil on the outside will ensure that the cheese does not become too dry and crumbly once it is removed from the humid conditions. It can be wrapped in cellophane or in cooking foil pierced with holes to allow it to 'breathe'.

White moulds   These develop on the outside of soft cheeses and the most famous

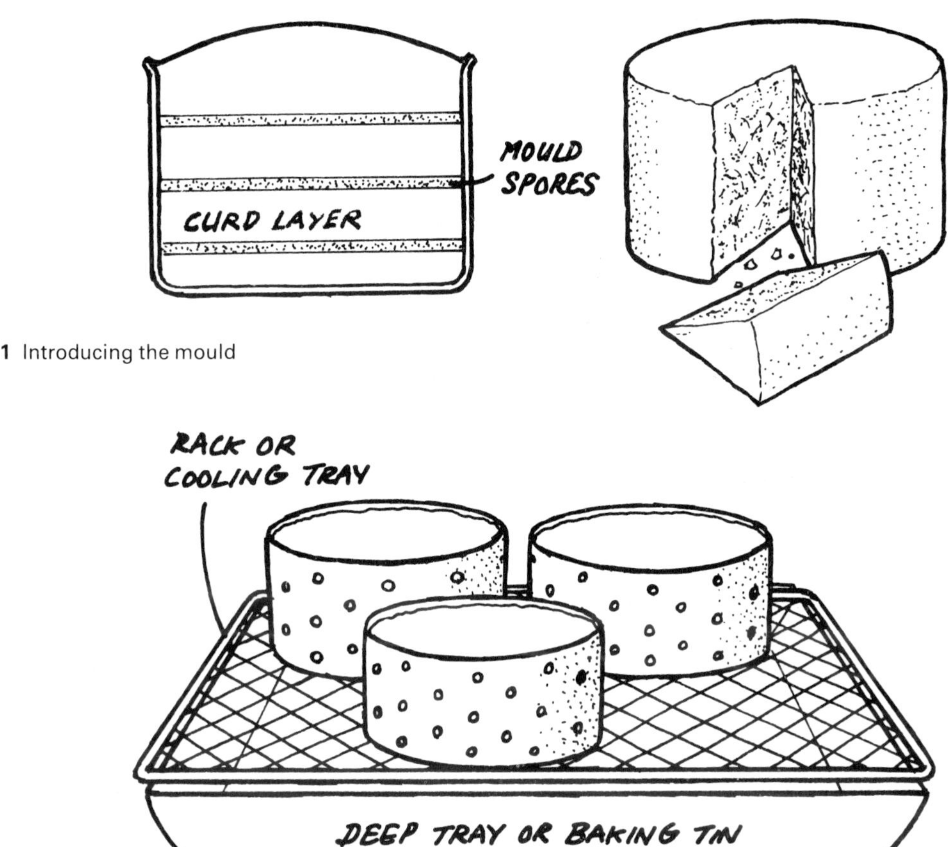

1 Introducing the mould

2 Draining the cheeses
3 *Below*, Making aeration holes

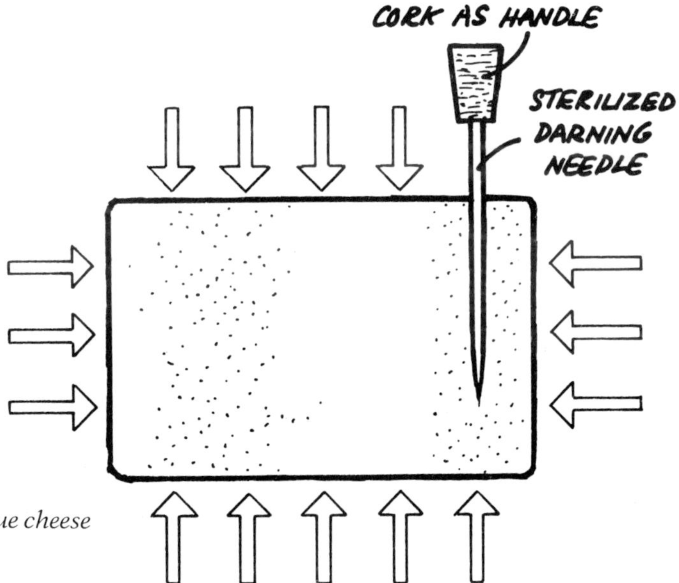

Fig. 25 *Making a blue cheese*

examples are Brie and Camembert. Spores of *Penicillium camemberti* and *Penicillium candidum* are available and these are normally sprayed on to the cheeses when they are put out to ripen. This can be a tricky technique, for an exceedingly fine spray is needed so that the cheeses do not become too wet. If the humidity is too high, blue or undesirable moulds may grow instead. The temperature for storage is also crucial. When first put out to ripen, the cheeses should be in a room at 15°C (60°F). A higher temperature will make them drain too quickly, producing a hard, dry cheese and inadequate mould growth. As soon as the white mould begins to appear they should be transferred to a room with a slightly lower temperature; 13°C (55°F) is ideal. The humidity must not be too high and the placing of a bowl of water as advised for the growth of blue moulds will be unnecessary.

On a small scale, using a home-made starter from a piece of shop-bought cheese will produce the correct 'aroma' and a certain amount of mould growth (see page 50), but better results will be had by the use of the commercial spores and it should be emphasized that for a really authentic white moulded cheese, the latter is essential. The ripening time will vary from two to six weeks, depending on the type of cheese.

Colouring cheese

Some cheeses are coloured to make them more yellow. Edam, for example, is bright yellow, while Red Leicester is a deep orange. The colouring does not affect the taste and it is a visual characteristic only. Soft cheeses made from goats' milk are pure white but it is not true, as some people claim, that hard cheeses made from goats' milk are also white. Goats' milk Cheddar, for example, is a very pale yellow and it is here that the addition of colouring may be thought necessary. For home production it is not really necessary and I rarely colour my cheeses.

Annatto from the plant *Bixa orellana* is used for colouring both cheeses and butter, but is prepared differently for the two uses. For butter the colour of the seeds is extracted in oil but for cheese it is prepared in water so that it mixes with fat and casein, and is in an alkaline solution. It is added after the starter is put in the milk and before the rennet is added. The amount needed is small and it is important to follow the manufacturers' instructions. The petals of Pot marigold (*Calendula officinalis*) were often used and it is worth experimenting with this. Take about half a cup of marigold petals, wash them, then put in a blender with a little hot water. Strain and add to the milk before the rennet is added. Make sure that it is the Pot marigold you use and not the French or South African marigolds which are often sold as bedding plants in early summer. Saffron has also been used as cheese colouring in the past.

Keeping records

If you are seriously interested in making cheeses other than at a superficial level, it is essential to keep a detailed record of your activities. It enables you to reproduce a particularly good cheese a second time and will also enable you to see where adaptations to such things as degree of acidity, duration of pitching and so on, may need to be made. If small-scale cheesemaking is to regain its traditional and rightful place in small farms and households it will do so partly because people will evolve their own methods through experience and the keeping of detailed records. Anyone can be lucky and produce a glorious cheese once, but it needs intelligence and application to do it a second time. Here is the record card which I produced for my own hard cheesemaking:

Pests and problems

Cheese in storage is vulnerable to pests and regular inspection is essential. The most prevalent are:

Cheese Record Card				
Date started:		Date pressing started:		Date pressing completed:
Date put to ripen:		Date ready for use:		
Quantity of milk used:		Amount starter used:		Amount rennet used:
Temperature at which starter added		Duration of scalding time		
Duration of time left before renneting		Duration of pitching time		
Acidity when rennet added		Acidity of whey after pitching		
Duration of time left before cutting curd		Duration of time for whey run-off or cheddaring		
Acidity of whey after cutting curd		Acidity before milling		
Temperature at scalding stage		Amount of salt added		
Details of pressure exerted (e.g. 1st day: 2 hours at 7 lbs. Left overnight at 28 lbs.)				
Type of Cheese	Texture		Taste	
Any other remarks				

Cheese fly   This is a small fly which lays its eggs in any convenient cracks in the cheese. The eggs hatch out in about three days and the larvae, known as 'jumpers', eat their way around the surface until they develop into adult flies in about three weeks. As their activities are normally confined to the surface it may be possible to salvage your cheese by cutting off the rind and outer sections.

Cheese mite   Much more serious are the depradations of the cheese mite which, strictly speaking, is not an insect but is allied to the spider family. It can do great damage to cheese in store and the eggs are difficult to eradicate because they are capable of withstanding extreme conditions. The eggs hatch in about ten days and the 'nymph' passes through two stages while it is developing. If the burrowing activities have substantially damaged your cheese there may be no alternative but to discard it.

House fly   This pest is so well known that it needs no description. It is one of the worst transmitters of disease-producing bacteria and the threat it poses as a health hazard to cheese is like that to any other food.

House mouse   Again, the activities of the house mouse need little description. It will gnaw considerable chunks out of stored cheeses, as well as leaving droppings around to soil them. Ideally a pest-proof room should be used, but if there is no alternative the placing of traps will be necessary. Most local authorities will also help and advise where there is a rodent problem of this kind.

The making of the cheeses themselves is not without its problems and the following are some of the more common ones.

Cheese too acidic   In this case too much lactic acid has been produced. Perhaps the original starter was added in too large a quantity, particularly if a home-made one was too acidic to start with. If the ripening period after the starter has been added is longer than that suggested it may also lead to too much acid. In the case of a harder cheese, leaving it too long in the whey after renneting or cutting the curd can produce this effect, or it may be the result of inadequate pressing so that too much whey is left behind. This whey will continue to become sharper as time goes on.

Cheese is rubbery If this is a soft cheese the cause is overheating or adding too much rennet, or both. Soft cheeses need only small amounts of rennet so that coagulation takes place slowly. Always use a dairy thermometer instead of trying to guess temperatures. In harder cheeses the cause may also be too much rennet or overheating, although some cheeses are required to have a rubbery texture. Edam is an example of this. Most hard cheeses will be rubbery when freshly made because the scalding stage requires the raising of the temperature, but the appropriate ripening period will affect the texture as well as the flavour.

Cheese tastes bad The most common reason for this is inadequate attention to sterilization techniques and general hygiene. It cannot be over-emphasized that milk products are ideal growing media for bacteria and cleanliness is essential.

Cheese has a fermented taste Yeasts are responsible for fermentation and have gained access to the curds. Make sure that dairy hygiene is practised and that no bread or winemaking is taking place at the same time as the cheesemaking.

Cheese has little flavour This may be due to inadequate ripening. See if there is an improvement after being left for a while longer. The cause may also be using an inadequate starter or not allowing the milk to ripen long enough for sufficient lactic acid to be produced.

Milk does not coagulate This may be the result of using dairy sterilizer and not rinsing the utensils properly so that it is killing off the lactic acid-producing bacteria in the starter. If rennet is used, perhaps not enough is being added or the temperature is not high enough. Another possibility, if annatto colouring is used is that it has been added at the wrong time. It should always be before the rennet is added, not afterwards when it can interfere with the curdling process.

Cheese is too dry and crumbly Too little rennet may cause this condition. Another possibility is that the curds have been cut into pieces which are too small or have been stirred too vigorously, leading to an excessive loss of the butterfat content.

Milk curdles instantly into small particles when rennet is added There is far too high a level of acid in the milk. Too much starter has been used or the milk itself may be almost sour.

Cheese is full of holes The cheese may be stored in temperatures which are too high, so that air in cracks is expanding. Cheese such as Gruyère which is required to have holes is stored in high temperatures for this reason. Another possibility is that yeasts have contaminated the curds and are producing carbon dioxide gas which is expanding.

An oily layer on the surface of the stored cheese The cheese is too hot and the fat is seeping out. Store at cooler temperatures.

Cheese is too moist Too much whey has been retained in the curds. Cut the curds into smaller pieces next time. Do not heat too rapidly. (In cheesemaking everything is done gradually.) In the case of a hard cheese the pressing may have been inadequate.

Cheesemaking in the tropics

The high temperatures in the tropics make the storage of cheeses difficult. The best ones to attempt are those which are highly salted such as the Feta cheese, or a soft cheese which is high in lactic acid such as the yoghurt cheese. Kefir grains can be used to produce yoghurt which is then strained and salted to produce a cheese. Where hard cheeses are attempted, they should be waxed rather than bandaged. This will offer more protection against 'sweating' of the cheeses where the butterfat oozes out, and also against drying up and hardening. The waxed cheeses can then be stored in a large bowl with a few inches of water to keep down the temperature. If this is changed at frequent intervals and the whole container covered with muslin the waxed cheeses should remain in good condition while they are ripening. Once they are cut they will need to be used up fairly quickly and kept in a refrigerator.

Selling cheeses

Cheeses made from cows' milk will come under the dairying regulations which are in force and further information should be obtained from the local department of agriculture. Goats' cheeses will need to meet the general food purity regulations and the Trades Description Act but are otherwise free of legislation. Most soft goats' cheeses are currently imported from the Continent and there is an almost totally unexploited market for good quality goats' cheeses in Britain and in many other countries. Ewes' milk cheeses also have a considerable potential.

Small cheese vats, moulds and other equipment are now available from several suppliers.

8

CHEESEMAKING RECIPES

This is an alphabetical list of recipes suitable for the home cheesemaker. Please note that the names of certain cheeses are correctly given only to cheeses manufactured under licence in certain areas, using specific techniques. Where I have used a specific name, it refers to a similar product but is not the same. For example, Brie should be interpreted as Brie-type and not the original Brie. Experimentation is a good thing and it should not be supposed that certain milk should only be used for certain cheeses. In recent years, too many books about cheeses have been written by people who have no practical experience and who have merely copied the printed errors of previous writers. For example, it is often claimed that goats' milk is only suitable for soft cheeses, presumably because the writers have only seen soft goats' cheeses offered for sale on the Continent of Europe and in delicatessens. Goats' milk can be used to produce a whole range of cheeses, both soft and pressed, including excellent Cheddar. The following recipes can therefore be tried with cows', goats' or sheep's milk. Relatively small quantities are given for some recipes so that you can produce small amounts of cheese at first.

Note: Use either metric, *or* imperial, *or* American measures in each recipe.

Appetitost cheese
1 litre (2 pints) (5 cups) fresh buttermilk

Leave the buttermilk to ripen for two days in a protected place, when the acidity should be approximately 0.4%. Put it in a saucepan and heat to 49°C (120°F) when the curds and whey will separate. Drain off the whey and leave the curds in a warm place covered with muslin for 48 hours. The natural enzymes will bring about natural fermentation of the milk sugar in these conditions. Heat to 49°C (120°F) again, by placing the curd in a bowl placed over a saucepan of hot water and work in salt to taste. Leave to cool in a refrigerator for a few hours when it will be ready to eat.

This cheese originated in Denmark and is so-named because it was claimed to improve the appetite. Try it as an *hors d'oeuvres* on strips of toast and see what happens.

Bondon cheese
2 litres (4 pints) (10 cups) fresh milk
125 ml (¼ pint) (⅝ cup) cultured buttermilk
3 drops rennet

Mix in the buttermilk and heat to 18°C (65°F), then add rennet. Leave to curdle overnight and the following morning drain and ladle the curds into a cloth. Hang to drain for two hours then place in a fresh cloth and put a scrubbed wooden board with a weight on top of the curds in the bound cloth. Leave overnight then sprinkle salt on the surfaces. Form into rounded cylinders to produce the traditional shape of a bung (from the French name 'bonde' meaning the bung of a barrel). This cheese originated in Normandy in the cider-producing area. Try it with crusty bread and cider.

Brie cheese
2 litres (4 pints) (10 cups) fresh milk
3 drops rennet
½ cup (1 tablespoon) of home-made starter or
* 1 teaspoon commercial starter*

This to some is the queen of cheeses. Its production is now a thriving French export business, a far cry from the small localized craft of its origin. In order to make a Brie-type cheese successfully it is essential to have a good starting culture. The easiest way of acquiring this is to use a small piece of shop-bought Brie and make up the starter in the way described on page 50. Once you have made your own Brie-type cheese you can use a piece of that as a starter for the next batch.

Warm the fresh milk to a temperature of 24°C (75°F) and stir in the starter. Leave for 15 minutes then add the rennet. If you are making cheese precisely and are able to test the acidity, it should be 0.21 but for small-scale production this is not essential. Leave to coagulate slowly for a few hours then cut the curd into cubes. Strain off the whey, ladle the curds gently into small moulds which are not too deep and leave to drain until the following day. By this time they will have settled to half their original volume and require firming. This should be done as carefully as possible so as not to damage the curds, and involves up-ending the mould on to a cheese mat. When the cheese is firm enough to be taken out of the mould – usually the following day – gently rub in salt on the surface and sides. A few hours later turn the cheese and rub in salt on the bottom.

During the next three days turn the cheese several times daily, each time rubbing in a little more salt. When firm, leave to ripen at a temperature of 15°C (60°F). Once white mould begins to appear put in a temperature of 13°C (55°F). A temperature higher than this should be avoided or the resulting cheese will be hard and dry. A much lower temperature will however encourage a crop of undesirable moulds on the surface. The best way of achieving the correct mould is to use a commercial preparation of *Penicillium candidum* or *Penicillium camemberti* available as spores. After about three weeks it is ready to eat. Do not leave it until it smells of ammonia for then it is too ripe and not fit to eat. Many authorities say that it is not possible to produce a Brie-type cheese under domestic conditions, but I have done it without too much difficulty. It is traditionally eaten with red Bordeaux wine or Médoc.

Buttermilk cheese
1 litre (2 pints) (5 cups) cultured buttermilk
½ cup of good mayonnaise
salt to taste

This recipe requires cultured buttermilk rather than that left over as a result of home buttermaking. It is available in cartons in many supermarkets.

Heat the buttermilk gently until the curds and whey are seen to separate. Leave for an hour then drain off the whey and leave the curds to hang in a muslin cloth. Leave to drain overnight then transfer the curds to a bowl and beat in the mayonnaise and salt. This cheese is nice with brown bread and beer.

Caerphilly cheese
5 litres (1 gallon) (20 cups) previous night's
* milking*
5 litres (1 gallon) (20 cups) morning's milking
½ cup (1 tablespoon) home-made starter or
* 60 ml (1 teaspoon) commercial starter*
* (necessary only if all-fresh milk is used)*
½ teaspoon rennet

This is a lovely cheese which originated in the town of Caerphilly in South Wales. It ripens more quickly than many other cheeses and also has a high yield (½ kg to 4.5 litres of milk or 1¼ lb to the gallon).

Skim the cream off the previous day's milking which has been allowed to stand overnight, then heat the cream to 32°C (90°F). Pass the cream through a strainer and add to the milk again. Add the morning's milk and after stirring thoroughly, raise the temperature of the whole to 21°C (70°F). Add starter at this temperature and gradually heat to 30°C (86°F) taking approximately 30 minutes to reach this temperature. Add rennet diluted in three times its volume of water and top stir to make sure that the cream is incorporated. Leave covered for 45 minutes to an hour when the curd will have set. The acidity of the whey should ideally be 0.22% at this stage. When the curd is ready, cut into cubes and stir gently with the hands for twenty minutes, and raise the temperature to 31°C (87°F). When the acidity of the whey has reached 0.52%, pour it off and break the curd into walnut-sized pieces. Unless you have a Lloyd's acidmeter to check the acidity, leave for a further 20 minutes when the approximate level of acidity will be reached. Add 1 level teaspoon of salt and ladle into a cloth in a mould. Apply pressure overnight then remove and immerse in a brine bath for 24 hours. The cloth should still be on the cheese. The strength of the brine is ½ kg salt to 4.5 litres of water (1 lb (2 cups) to 1 gallon). Leave to drain for 24 hours in a cool place then allow to ripen for ten days, turning it twice a day. Caerphilly goes well with a dry red wine such as a Médoc.

Cambridge cheese (also known as York)
7 litres (1½ gallons) (30 cups) full cream milk
1 teaspoon commercial starter or 1 tablespoon cultured buttermilk
½ teaspoon rennet

This is a mildly acid soft cheese with a characteristic orange stripe. It was traditionally made in the Cambridge area and sold in local markets but has now disappeared except for those made by home cheesemakers. Heat the milk to 68°C (155°F) and cool immediately to 30°C (86°F). Add the starter and stir well. Leave to ripen for 10 minutes. Dilute rennet in three times its own volume of water, and add to the milk.

Stir in well then transfer 2 litres (4 pints) (10 cups) of this to a sterilized bowl. Add a few drops of cheese annatto until the mixture is pale orange. You can produce your own preferred shade of orange by adjusting the number of drops.

Leave to set for 30 minutes, and meanwhile boil cheese mats, bottomless moulds and trays for 10 minutes. Stand the moulds on the mats in the trays and ladle thin slices of the white curd into two moulds until just over a third full. Divide the orange curd between the two moulds and then finish off with white curd up to the top. Cover with greaseproof paper and leave to drain for 2–3 days until the cheese is about 5 cm (2 in) high and firm. Remove from the moulds and lightly salt. Wrap carefully in greaseproof paper and put in the refrigerator. It is ready to eat in 12 hours and should be eaten within a week.

A light medium sweet wine makes a good accompaniment to Cambridge cheese.

Cantal cheese
5 litres (1 gallon) (20 cups) previous night's milk
5 litres (1 gallon) (20 cups) morning's milk
1 teaspoon commercial starter (necessary only if all-fresh milk is used)
1 teaspoon rennet
Salt
Brine

This is an ancient pressed cheese traditionally made in the Auvergne and sometimes referred to as the 'French Cheddar'.

Leave milk overnight and in the morning add the morning's milk. Heat to 49°C (120°F) then cool to 30°C (85°F). Add starter and stir thoroughly. Leave covered overnight and then add rennet. When the curd is firm cut into cubes and stir with the hands. Heat for half an hour at 38°C (100°F). Leave for half an hour without heat, then drain off the whey. Place the curds in cheesecloth and hang up to drain. Sprinkle salt at the rate of 28 gm to 1.3 kg curd (1 oz to 3 lbs) (2 tablespoons to 8 cups) and place in a mould in a cloth. Put in a press for 24 hours, then remove, turn upside down and replace under pressure. After another 24 hours remove the cheese and turn upside down again. Leave for 48 hours. At the end of this time remove the cheese and immerse in brine for 5 minutes. The brine should be approximately 4 tablespoonfuls (6 tablespoons) salt to 1 litre (2 pints) (5 cups) water. Replace in the press. Repeat this twice at 48 hour intervals, then bandage the cheese for storage. It should be stored at 10°C (50°F) for between three to six months in order to ripen properly. Turn it every day for the first week, then once every three or four days after that.

A good full-bodied red wine is the traditional accompaniment to Cantal but try it with good English beer and pickled onions.

Caws bach (Little cheese)

2 litres (¹/₂ gallon) (10 cups) raw milk
300 ml (¹/₂ pint) (1¹/₄ cups) natural buttermilk

This is a traditional Welsh cheese and is one of my mother's recipes. Slowly heat half a gallon of raw milk to 24°C (75°F) and then stir in half a pint of nicely soured buttermilk. Stir well and leave for 24 hours in a warm place. When the curd is firm, ladle into cloths and tie very tightly. Leave to hang for 24 hours, but twice during that period, undo the cloths and scrape the drier curd into the softer inner curd. Add salt to taste, and either finely chopped chives or spring onions.

Cheddar cheese

10 litres (2 gallons) (40 cups) fresh full cream
  milk
the cream from 10 litres (2 gallons) (40 cups)
  of the previous night's milking
¹/₂ cup (1 tablespoon) home-made starter or 1
  teaspoon commercial starter (necessary only
  if all-fresh milk is used)
1 teaspoon rennet
salt

It is fashionable to be disparaging about Cheddar cheese with dismissive comments about its suitability for mouse-traps, but in my view, it is one of the great cheeses of the world. No-one seems to be able to make it as well as the British and the so-called. Cheddar that I have sampled in the USA tasted of bland pasteurization and nothing else. Even the French with their deserved reputation for superlative soft cheeses, make dreadful Cheddar. Here is the traditional way of making it before pasteurization became the norm. If you wish to pasteurize your milk at the start, heat it to 68°C (155°F), otherwise proceed as follows:

Strain and cool the evening's milk and leave to stand overnight. In the morning skim off the cream, warm, strain and mix it with the morning's milk, discarding the skimmed milk for other purposes. Add the starter and leave for about thirty minutes when the acidity of the milk should ideally be between 0.18–0.20%. Heat the milk to 30°C (86°F) and add rennet. Stir thoroughly making sure that the cream at the top is also stirred in. Leave for ten minutes then stir the surface again to prevent the cream rising to the top. Leave until the curd is firm and breaks cleanly and no milk stain is left on the back of the finger when tested. Do not leave it for longer than this unless you particularly want a dry, crumbly cheese. It normally takes about 45 minutes to reach this stage. Cut carefully into cubes and stir gently with the hand for a few minutes. Gradually heat so that the temperature rises slowly over a period of 45 minutes to 38°C (100°F). Continue stirring by hand while this is going on. Remove from the heat and pitch or swirl the whey with the hand so that the curds settle in a heap at the bottom. Drain off as much whey as possible and leave the curds to drain in a cloth until they have all formed a single mass. Cut the curd into large strips and place one on top of the other in cheese cloth on a draining mat or tray and leave for 15 minutes. Rearrange the order of the strips so that the outer ones are in the middle (see Fig. 23). Leave for another 15 minutes until all the gas holes have appeared and been dispersed. At this stage the curd is ready for passing through a curd mill, slicing into thin flakes with a sharp knife or breaking into small pieces by hand. Add salt at the rate of 28 gm to 1.3 kg (1 oz to 3 lbs) (2 tablespoons to 8 cups). The traditional way of doing this in the farmhouses was for two dairy maids to hold the cloth, one at each

Fig. 26 *A range of cheese moulds for hard and soft cheeses*

end and toss the curds so that the salt was mixed in well. (History does not relate what happened if any of the curds fell on the floor.) The optimum acidity at this point should be 0.60%.

Pack the curds into moulds lined with sterilized cloths and apply a light weight for the first hour. Increase the weight by 50% for the next hour, then increase again to maximum pressure. This gradual pressure is to avoid squeezing out the fat which would harden on the surface and impede drainage. Leave for 24 hours, take out of the press, replace in clean cloths and put back, upside down, in the moulds. Exert full pressure and leave for another 24 hours. Remove from the press and leave to dry for a few hours, then bandage firmly (see page 59). Leave the cheese to ripen in a cool, dry place where it should be turned daily for the first three weeks, then on alternate days after that. For a mild cheese, ripening should take place for between 3–5 months.

For a mature cheese, leave it for at least 5 months but preferably for 6–7 months.

Cheshire cheese

5 litres (1 gallon) (20 cups) previous evening's milk
5 litres (1 gallon) (20 cups) morning's milk
½ cup home-made starter or 1 teaspoon commercial starter (necessary only if all fresh milk is used)
½ teaspoon rennet

The Cheshire is an ancient cheese with the distinction of having been mentioned in the Domesday Book, although folklore claims that it was in Britain before the Romans. It is a friable, salty cheese useful for cooking as well as eating.

Strain and cool the evening's milk and leave to stand overnight. Stir at intervals to prevent cream rising to the surface. Any cream which does rise should be skimmed off in the morning, warmed then mixed with the morning's milk. Combine the two milks, then heat to 49°C (120°F) and cool to 30°C (86°F). Add starter and stir thoroughly. Leave until the acidity has reached 0.20%. If an acidmeter is not available, leave for 1½ hours. Heat to 30°C (86°F) and add rennet. Stir for ten minutes to mix and to disperse the cream, then leave to coagulate. When the curd is firm and breaks cleanly, cut into 3 cm (1 in) cubes, being very careful not to handle it roughly or the fat will be lost in the whey.

Leave for five minutes, then heat gradually to 34°C (93°F), stirring gently with the hands all the time. Remove from heat and leave to settle for 40 minutes. Drain off the whey and cut curd into 15 cm (6 in) cubes. Place on cheese cloths spread on trays and after 15 minutes turn upside down. Turn again 15 minutes later, then again after the same interval of time. Break into pieces the size of a bean and add salt at the rate of 28 gm to 1.3 kg curd (1 oz to 3 lbs) (2 tablespoons to 8 cups). Pile into lined moulds and place in a temperature of 24°C (75°F) for 24

hours. (An airing cupboard is suitable – or on a rack above the Aga, but do ensure that the cheese is kept covered.) After 24 hours place in clean cloths in the moulds and apply pressure: lightly at first and gradually increasing it over the next 12 hours. Leave for 24 hours then upturn and press for another day. After three days, remove and bandage. Ripen at a temperature between 13°C–15°C (55°F–60°F) and turn frequently. It is ready at 3 months. Beaujolais is a nice wine to drink with Cheshire cheese but the traditional beverage is good ale.

Club cheese

This is prepared from any stale, left-over cheese. Grind it up then blend with a little butter. And a small amount of Roquefort or Danish Blue cheese and season with pepper. Pack into small glass jars or wrap in foil until used.

Colby cheese

5 litres (1 gallon) (20 cups) fresh milk
1 tablespoon commercial starter or 1 tablespoon fresh buttermilk
1 teaspoon rennet

This cheese was traditionally made from fresh milk straight from the cow while it was still warm. This is my mother's recipe.

Add the starter to the lukewarm milk and leave covered overnight. In the morning heat to 27°C (80°F) and add the rennet which has been previously diluted with three teaspoons of water. Leave to curdle until it is firm to the touch. Cut the curd into cubes and stir carefully to release the whey. Heat to 38°C (100°F) and maintain at this temperature for half an hour. Stir gently throughout this period. Turn off the heat and leave to stand for an hour stirring occasionally. Drain off the whey through a cloth and leave the curds suspended to drain for a further half an hour. Add salt to taste but do not break up the curd. Place carefully into a mould and press for two days, gradually increasing the weight every 12 hours. Turn the cheese upside down after the first day. Remove from the mould and leave to dry for two days then wax or bandage and ripen for three months. This cheese was often coloured with annatto, but it is not necessary. It is apparently an American cheese, but I do not know where my mother got the recipe, although my father's father was an American sailor so that may have something to do with it.

Colwick cheese

5 litres (1 gallon) (20 cups) full cream milk
1 teaspoon commercial starter or 1 dessert-
spoon cultured buttermilk
½ teaspoon rennet

This is a smooth, slightly acid curd cheese, normally served with clotted or whipped cream in the hollow centre – so it is definitely not for calorie watchers. A tallish mould such as the coulommier mould is the most suitable, but the cheese is not turned at all. Heat the milk to 68°C (155°F) and cool immediately to 30°C (86°F). Add the starter and then ½ teaspoon of rennet, diluted in three times its own volume of water and stir thoroughly. Continue to top stir to prevent the cream floating on the top until curdling begins then leave to coagulate slowly for 40 minutes. Line the moulds with previously boiled muslin and ladle slices of curd into them. After an hour, pull the muslin inwards and upwards, thus drawing the curd away from the sides of the mould, then tie firmly. Repeat this at hourly intervals, until the cheese has a hollow in the middle, with the edges curving inwards. When the cheese is firm enough to handle, usually after two days, take it out of the mould and peel off the muslin. Sprinkle lightly with salt and just before serving fill the hollow with clotted or whipped cream. It should be eaten within three days. A very light and dry white wine goes well with this cheese.

Cottage cheese

See skimmed milk cottage cheese.

Coulommier cheese

3 litres (6 pints) (15 cups) milk
1 teaspoon commercial starter or ¼ cup (½
tablespoon) starter made from Brie cheese
3 drops rennet

This is a mild French soft cheese, similar to an unripened Brie. Purpose-made stainless-steel Coulommiers moulds are available or two plastic, open-ended moulds one on top of the other and secured with sticky tape will suffice. Heat the milk to 68°C (155°F) and cool immediately to 30°C (86°F). Add 1 teaspoonful of starter and stir thoroughly. If commercial starter is not available, use a piece of shop-bought Brie to make your own (see page 50). Leave for 20 minutes then add 3 drops of cheese rennet which has been previously diluted in 4 teaspoonfuls of warm water. Stir and leave for half an hour, or until the curd is firm, and does not leave a milk stain on the back of your finger.

Meanwhile, sterilize two cheese mats by boiling for a few minutes. Put the moulds on a cheese mat and ladle the curd into the moulds until full. Leave until the curd has sunk to below the collar mark where the two halves· of the mould interlock, and remove the top half. This shrinking will take several hours, and during this time keep the moulds covered with the second mat. Turn the cheese onto the second mat, remembering to sterilize the mat first. By the following day, the curd should have shrunk to half way down the mould, and will be firm enough to remove. Sprinkle salt on the top and bottom. Keep in the refrigerator for a few hours, when it will be ready to eat, but develops more of a flavour if allowed to ripen for three days. The traditional wine accompaniment is Nuits St George.

Fig. 27 *Making a Coulommier cheese*

Cream cheese – single
1 litre (2 pints) (5 cups) fresh single cream
  (light cream)
1 teaspoon commercial starter
3 drops rennet

This is a soft, granular and rather buttery cheese which is unripened. Heat treat the cream to 80°C (175°F) in a double boiler or basin on top of a saucepan of water. Cool immediately to 24°C (75°F) by placing the bowl in cold water. Add the starter. Cover and leave for three hours. Stir in the rennet diluted in six times its volume of previously boiled and cooled water, cover and leave to coagulate for eight hours. When the curds are thick, ladle into a boiled and cooled cloth and hang to drain in a cool pantry or dairy. A temperature range of 10°C–13°C (50°F–55°F) is ideal. Next day open up the cloth and scrape the curd from the outside to the inside so that draining and drying takes place throughout the curd. Leave to drain for another day then sprinkle in salt to taste. Package in small plastic or metal foil containers and store in the refrigerator until needed. It is ready to eat straight away but will keep for up to a week under refrigeration.

Cream cheese – double
Use double cream (heavy cream) instead of single (light) and prepare in the same way, but omitting the rennet.

Crowdie cheese
I am indebted to Mrs G. Mackintosh for the recipe for this Scottish cheese which was traditionally eaten for breakfast with oatcakes and butter.

2 litres (4 pints) (10 cups) milk
300 ml (½ pint) (1¼ cups) double (heavy)
  cream
½ teaspoon rennet
Salt to taste

Warm the milk until it is barely lukewarm then add the rennet and leave for three hours. Cut the curd into cubes and leave in the whey for another three hours. Drain the curds through muslin and add the cream and salt. Beat to a thick paste and then place in containers. Refrigerate to cool and firm the cheese then it is ready for eating straight away. If you want to make a traditional Highland version of this cheese, add 55 gm (2 oz) (4 tablespoons) butter at the beating stage.

Curd cheese
See lactic curd goat's cheese.

Derby cheese
I am indebted to Mrs Hunt of Derby for this recipe.

5 litres (1 gallon) (20 cups) previous night's
  milk
5 litres (1 gallon) (20 cups) morning's milk
½ teaspoon rennet

This is a nice cheese with a flaky texture and originated in the county of the same name. Commercially it is sold after ripening for one month and is far too mild and bland. Traditionally it was ripened for at least two months when a fuller flavour developed. It is only home cheesemakers who now produce the real thing.

Mix the morning's milk with the previous evening's which will have ripened overnight. Heat slowly to 29°C (84°F) taking about 30 minutes to reach this temperature. By this time the acidity will be between 0.19–0.21%. Add rennet and stir. Leave for 45 minutes then cut the curd into cubes. Heat to 35°C (94°F) stirring the curds in the whey by hand. Stop heating and leave the curds to settle for 30 minutes. Drain the whey and leave the curd in a cloth which has the knot gradually tightened once every quarter of an hour for an hour. Remove from the cloth and cut the curd into four wide strips, piling them on top of each other. Reverse the order after half an hour and leave for a further 30 minutes. Cut into small pieces and add salt at the rate of 28 gm to 1.9 kg curd (1 oz to 4 lbs) (2 tablespoons to 11 cups). Put into lined moulds and exert light pressure for an hour, gradually increasing it over the next four hours. Leave at maximum pressure for 24 hours then remove, upturn and replace in a clean cloth under form pressure. Remove after two days and rinse in weak brine. Bandage and store for two months before eating, although if you cannot wait that long, eat it after a month. Derby cheese is traditionally eaten with pickled onions, soft rolls and light ale.

Devon Farmhouse cheese

I am indebted to Mrs Wheeler of Dunchideock in Devon for this recipe.

5 litres (1 gallon) (20 cups) previous evening's milk.
5 litres (1 gallon) (20 cups) morning's milk
60 ml (½ teaspoon) commercial starter
½ teaspoon rennet

Mix the milks and heat to 32°C (90°F) by standing the pail in hot water and stirring. Now add starter and stir it well. Cover the container with a clean cloth and leave for 45 minutes. Now add ½ teaspoonful rennet, diluted with 2 teaspoonfuls cold water, by pouring it over the perforated skimming ladle, thus strewing it into the milk and stir it well right down to the bottom of the pail for at least 1 minute, then topstir with the flat underside of the ladle not more than ½ inch deep, for three minutes. Cover container and leave for 45 minutes when the curd will have formed nicely.

With a long knife or palette knife, cut the curd at ½ inch intervals, then at right angles again cut it across and across. Using the ladle cut spirally downwards, starting in the middle at the top. Now turn the curds right over, cutting up any large ones, and continue this stirring for 30 minutes, bringing the temperature slowly up to 38°C (100°F). Cover the container again for a few minutes to allow the curds to settle, pour off the whey and gather the curds into a scalded and wrung-out butter muslin square, using one corner to tie around the other three, thus tightening the bundle for more whey to drain out. Leave it for about one hour, tightening it every now and then.

Have the cheese press ready. Scald the tray, mould, round follower and muslin. Line the mould with the muslin. Tip the curd into a scalded dish and break it up with your fingers, gently but firmly, into walnut sized pieces and add block salt at the rate of 28 gm per 1.8 kg curd (1 oz per 4 lb). Mix well. Pack the curds into the mould, fold the muslin neatly over the top, put on the round follower and put under 20 lb pressure for 2–3 hours. Now turn the mould upside down, put on the follower again and increase pressure to 30 lb for another 2 hrs. Increase pressure to 40 lb and leave until next day. Now turn the mould again and put under 50 lb pressure for 24 hours or longer.

Remove the cheese from the mould and muslin. Leave it in a warm and airy place for 2–3 days, turning it frequently to air-dry the outside, then dip it in warm paraffin (cheese) wax to coat all over. Put it to ripen in a cool store, turning it only once a week. It can be eaten after five weeks but improves with keeping.

Dorset Blue cheese

5 litres (1 gallon) (20 cups) skimmed milk
5 litres (1 gallon) (20 cups) fresh morning's milk
½ teaspoon rennet

Strain and cool the evening's milk and the following morning skim off the cream. Put the cream to one side for other use, such as buttermaking. Add the morning's milk to the skimmed milk and add the starter. Leave for 45 minutes then heat to 30°C (84°F) and add rennet. Leave for one hour then cut the curd into cubes. Leave for 15 minutes, then heat to 32°C (90°F) and stir continually for 1 hour. Drain off the whey and cut the curd into squares. Leave them on cloths on a rack for an hour. Add salt at the rate of 56 gm to 2.25 kg of curd (2 oz to 5 lbs) (4 tablespoons to 13½ cups). If you have a commercial sachet of freeze-dried *Penicillium roqueforti* add the spores at the 'starting' stage. Ladle into lined moulds and put under pressure. Gradually increase the pressure over the next three days, turning the cheese in the mould daily. After three days, store in a cold room and pierce the cheese with a sterilized stainless steel needle which has been first dipped in blue mould from a cheese such as Roquefort, Blue Stilton or Danish Blue unless the commercial mould was used. The blue mould should begin to spread after about a week, but you may need to pierce the cheese several times again to ensure that sufficient air is getting in.

Double Gloucester cheese

I am indebted to Mrs Jean May for this recipe.

5 litres (1 gallon) (20 cups) fresh, full cream milk
½ litre (1 pint) (2½ cups) raw cream left to ripen for two days
½ teaspoon rennet
2 teaspoons annatto colouring (optional)

Mix in the well-ripened cream which acts as a starter as well as providing extra creaminess. (It should not be forgotten that the original cheese was made from the summer milk of Gloucester cattle which had a very high butterfat content.) Heat slowly to 30°C (86°F), taking about fifteen minutes to reach this temperature when the acidity will be about 0.18%. Leave for ten minutes then add the rennet, swirling it into the milk and stirring in the top cream at the same time. The temperature should be held at the previous 30°C (86°F) and this is where a small vat with a surrounding water bath comes in useful for there is less heat loss during the waiting periods. In 40 minutes the curd should be firm and ready to cut. Cut into cubes in the same way as for Cheddar, then stir them by hand while slowly raising the temperature of the whey to 37°C (98°F), taking half an hour to reach

this temperature. Do not let the temperature go higher than this. Leave the curds to settle in the whey for ten minutes then drain into a cloth and leave for half an hour. Undo the cloth and take out the curd mass. Cut into 10 cm (4 in) cubes and stand on mats to drain for half an hour. Turn them over twice during this time. Now break into small pieces and add salt at the rate of 28 gm to 1.3 kg curd (1 oz to 3 lbs) (2 tablespoons to 3 lbs). Put into lined moulds and press, gradually increasing the pressure every hour. Remove the cheese from the mould after two days and dip in hot water for a few minutes then leave to dry for a week, turning daily. After a week bandage or wax. If a coloured cheese is required, add the colouring before the rennet.

Dunlop cheese

5 litres (1 gallon) (20 cups) previous night's milking
5 litres (1 gallon) (20 cups) morning's milk
2 teaspoons (2½ teaspoons) rennet

This is a Scottish cheese which is now only made in factories. To my knowledge I am the only one who makes it in the traditional way (although I hope that I am wrong).

Skim the cream off the previous night's milk and after warming it mix it in with the two lots of milk, skimmed and unskimmed. Heat to 30°C (86°F) and add the rennet which is undiluted. Stir well and it will have curdled into a firm curd within a quarter of an hour. Cut into cubes. From this point on follow the directions given for Cheshire cheese.

Edam cheese

5 litres (1 gallon) (20 cups) milk
3 teaspoons (3½ teaspoons) rennet
1 teaspoon annatto cheese colour

Heat the milk to 34°C (93°F) and add cheese annatto for colouring. Add the rennet and cut the curd into cubes after 15 minutes.

Heat the curd to 35°C (94°F) stirring continuously, then drain. Ladle the curds into lined moulds while still warm and apply pressure. After 24 hours remove, turn upside down and replace in clean cloths in the mould. Apply pressure for another 24 hours, then remove and dip in hot whey which has been heated to 47°C (126°F). Press again for 12 hours then place in a brine bath for 5 days; ¼ cup salt to every litre (2 pints) of water.

Remove, dry and wax the cheese. Allow to ripen for three months before use.

Feta cheese

5 litres (1 gallon) (20 cups) ewe's milk or goats' milk to which ½ litre (1 pint) (2½ cups) of cream has been added. Anglo-Nubian goats' milk will not require cream.
1 cup raw ewe's or goat's milk ripened for 36 hours or 1 teaspoon commercial starter
½ teaspoon rennet
Salt

This is a soft, salty cheese which originated in Greece, but is widely made in Bulgaria, Yugoslavia and other parts of Europe where flocks of dairy ewes have been kept for many generations.

Add the cup of ripened milk to the larger quantity of milk and stir well to mix and to incorporate the cream which tends to float to the top. Heat slowly until a temperature of 29°C (85°F) is reached in 30 minutes. Add the rennet which has been previously diluted in four times its volume of water. Mix well, remembering to top-stir again to keep the cream mixed. Cover and leave to curdle for half an hour. Cut the curd into 3 cm (1 in) cubes and leave in the whey for a further 15 minutes. Now stir the cubes gently in the whey for a further 15 minutes. Line a large colander with muslin and drain the curds of whey. Leave the curd mass to continue draining like this for about three hours, then lift up the cloth and upend the curd mass onto a second piece of muslin to continue draining for another hour. Cut the curd into 8 cm (3 in) blocks and sprinkle them with a little salt on both sides. Leave

to continue draining and drying on mats for three days, rubbing in a little salt each day. They are ready for eating after this but, if preferred, can be placed in a plastic box and covered with brine made up of two tablespoons of salt in 1¼ litres (2 pints) of water and stored for several weeks.

Dry Greek wines, ouzo and retsina go well with this sharp, salty cheese.

French Goats' cheese

5 litres (1 gallon) (20 cups) fresh, full cream goats' milk
½ cup (1 tablespoon) raw goats' milk left to ripen for 24 hours before use
3 drops rennet

France is justifiably famous for its goats' cheeses but many of them are so easy to make that it is surprising that most people pay high prices for imported ones when they could so easily make their own.

Add the ripened milk to the fresh milk; the former acts a starter. Heat the whole lot of milk to a temperature of 22°C (72°F). Dilute the three drops of rennet in a tablespoon of boiled and cooled water and stir in. Cover the pan and leave to coagulate overnight. The following day ladle the curds into small plastic moulds with bottoms; colwick moulds are ideal, or plastic yoghurt pots pierced with holes can be used. Leave them to drain on a rack placed over a baking tin. Pile the curd in because over the next two days it will sink to about half its height. After two days remove the cheeses from the moulds rub a little salt on the surfaces and leave to dry on mats. Once they are dry, after about 24 hours, they can be packaged in cellophane paper or foil pricked with holes to let the cheese 'breathe'.

Gervais cheese

5 litres (1 gallon) (20 cups) whole cream milk
2½ litres (4 pints) (10 cups) cream
¼ teaspoon rennet

Mix the milk and cream, stirring well, while slowly raising the temperature to 18°C (63°F). Add rennet, previously diluted with ten times its volume of boiled and cooled water. Cover the milk and leave to stand until the following day. Cut the curd just enough to make ladling possible and ladle carefully into a boiled cloth. Hang the cloth to drain. After three hours take it down and scrape the curd off the cloth and repack so that the outer part is on the inside, and vice versa. When the curd is fairly firm add salt to taste and put in small moulds lined with clean, white blotting paper which is an excellent way of extracting the remaining liquid without losing the fat. It is essential not to apply any pressure, for this will result in a loss of fat. After two days remove the cheeses from the moulds and leave on a mat for another few hours before packing or refrigerating. The cheese is ready for eating straight away or may be left to mature for three to four days.

Gorgonzola cheese

5 litres (1 gallon) (20 cups) full cream milk
½ litre (1 pint) (2½ cups) cream (optional)
1 teaspoon commerical starter or 1 dessert-
spoon home-made starter
½ teaspoon rennet

This originated in the Italian village of that name, although its production has now spread to many other countries. It is essential to have a pure culture starter and this may be obtained by using a sample from a shop-bought Gorgonzola (see page 50 for instructions on how to prepare it). The blue mould can also be introduced in this way, or by the use of commercial freeze-dried spores of *Penicillium roqueforti*.

Full cream cow's milk should be used, or the milk of Anglo-Nubian goats. If other goats' milk is used, then the addition of extra cream will be required. Heat the milk to 26°C (80°F), add the starter culture and stir in well. Leave the milk for half an hour when the acidity level of 0.27% is reached. Bring back the temperature to 26°C (80°F) and add rennet. When the curd is firm, after about 45 minutes cut into cubes and put into moulds which are perforated at sides and bottom, alternating the curd layers with a thin sprinkling of the famous mould scraped off a shop-bought cheese. Leave to drain and turn the moulds every day. Alternatively, use the commercial spores referred to above and introduced at the 'starting' stage. Remove from the moulds when firm and rub in salt on the surface. Leave to ripen for four weeks at a temperature of 10°C (50°F). After this period rub on a light coating of vegetable oil to prevent shrinkage.

When the cheese is ripening, a slimy mould will probably develop on the surface. This should be scraped off and holes made with a stainless steel skewer, to allow the air to enter, otherwise the interior mould cannot develop. Be prepared for failure because this is a difficult cheese for the amateur to produce.

Gouda cheese

I am indebted to Mrs Wheeler of Dunchideock in Devon for this recipe.

5 litres (1 gallon) (20 cups) full cream milk
from evening's milking
5 litres (1 gallon) (20 cups) full cream milk
from morning's milking
1 teaspoon starter
1 teaspoon rennet
Cheesewax

Hear the milk quickly to 66°C (150°F), then cool quickly to 32°C (90°F). Add the starter, stir well and add 1 teaspoonful rennet, diluted with 3 teaspoonfuls cold water. Deep stir for one minute then top stir for three minutes, cover and leave for an hour. Cut curds and then take 30 minutes to heat to 38°C (100°F), stirring all the time. Continue to stir at this temperature for a further 30 minutes, and during this time take out 2½ litres (4 pints) whey at a time, replacing it at once with the same amount of water at the same temperature and do this three or four times. This gives the cheese the typical smooth texture. Now pour off all the watery whey, allow the curd to mat into one lump. Have the mould ready lined with muslin and pack the curd into it, breaking it as little as possible, fold over the muslin, add the round follower and put under 20 lb pressure for 20 minutes. Turn the mould over, and put under 30 lb pressure for 20 minutes. Turn again, increase pressure to 40 lb and leave it for 3 hours, longer for a larger cheese. Prepare a 20% brine by mixing 570 gm block salt in 2½ litres cold water (1¼ lbs in ½ gallon) (5½ cups in ½ gallon) and float the cheese in this for 3 hours. Take it out, mop it dry, put to ripen at 10°C (50°F) for 3 weeks, rubbing it with a dry cloth daily and turning it. It can then be waxed to be left to mature for longer.

Gruyère cheese

*2½ litres (4 pints) (10 cups) evening's milk
 that has been skimmed*
2½ litres (4 pints) (10 cups) morning's milk
½ teaspoon rennet

Skim the cream from the evening's milk and use it for something else. Mix the two milks and gradually heat to 92°F (33°F). Add the rennet when it is at this temperature, stirring it in well. Leave to curdle until the curd is firm and breaks cleanly. Heat gradually to 60°C (140°F) stirring all the time so that the curd is broken up into particles the size of wheat. Drain and ladle the curds into cloths, adding salt at the rate of 28 gm to 800 gm curd (10 oz to 1½ lbs) (2 tablespoons to 4 cups). Put into lined moulds and press lightly overnight. The following day remove and dip in brine. Leave to dry and cure on mats at a temperature of 21°C (70°F) which is high enough to produce carbon dioxide in the curd. As this expands it produces the characteristic holes in the cheese. Leave to ripen for at least three weeks before use but if it still lacks flavour, leave for another few weeks.

Lactic curd goat's cheese

10 litres (2 gallons) (40 cups) fresh milk
*90 ml (3 fl oz) (6 tablespoons) commercial
 starter*
½ teaspoon rennet
1 teaspoon salt

Heat the fresh milk to 71°C (160°F) then cool to 24°C (75°F). Add the starter, stirring it in well and leave covered for two hours. Stir in the rennet, cover and leave for 24 hours. Next day stir the curds to break them up and drain off the whey by pouring through a cloth into a sterilized bucket or large bowl. Hang the cloth to allow the curds to drain. After two hours untie the bundle and scrape the outer curd into the middle and vice versa. Sprinkle on the salt and leave to drain for another few hours. If necessary, place a scrubbed beech board on the bundle with a weight on top. The following day it should be ready to package in plastic containers. This cheese is probably the best one to make if you wish to sell goats' cheeses. It is essential to use a commercial starter if you do so and to label each container, 'Lactic curd goat's cheese.'

Leicester cheese

5 litres (1 gallon) (20 cups) evening's milk
5 litres (1 gallon) (20 cups) morning's milk
*2 teaspoons (2½ teaspoons) annatto cheese
 colour*
*1 teaspoon commercial starter (necessary
 only if all-fresh milk is used)*
1 teaspoon rennet

Leicester originated in the Rugby area of England and is reddish in colour; hence its alternative name of Red Leicester. Add the morning's milk to the previous evening's and heat to 30°C (86°F). Add starter and annatto colouring. Add rennet then leave for 45 minutes to an hour when the acidity should be approximately 0.20%. Traditionally the curd was cut in a circular direction, starting from the outside of the circular vat, in a spiral towards the middle.

Then it was cut across and finally horizontally. When cut the curds should resemble peas. Heat to 35°C (94°F), taking an hour to reach this temperature. Drain off the whey and cut the curd into wide strips which are piled one on top of another. Turn them at intervals and after an hour add 28 gm salt to 1.3 kg curd (10 oz to 3 lbs) (2 tablespoons to 8 cups). Put in lined mould under slight pressure for an hour, then turn and increase pressure. Leave for 24 hours under maximum pressure. Bandage and turn daily. It is ready in three months.

Lemon cottage cheese
I am endebted to Rosemary Dick of Helston, Cornwall, for this recipe.

2½ litres (4 pints) (10 cups) milk
Squeeze of lemon juice
Few drops rennet

Heat the milk to just below boiling point and add a squeeze of lemon juice. Leave until blood heat, then add a few drops of cheese rennet. Leave for a few hours to set firm, then ladle curds into a large strainer or sieve. Stir now and again while the whey drains off. (The whey can be fed to poultry.) Add a little seasalt to taste. This cheese is nicest used really fresh, and if you add a little chopped celery and some walnuts, and serve it with salad it makes a really nice meal. Chopped chives or almost any herb you like can be used.

Little Welsh cheese
This should not be confused with Caws bach (Little cheese) (see page 69) which is also a Welsh recipe.

5 litres (1 gallon) (20 cups) fresh milk
½ teaspoon of rennet
6 sprigs of parsley
Salt to taste

Leave the fresh milk to ripen overnight then warm slowly to a temperature of 29°C (85°F). Add the rennet dissolved in three times its volume of water. Stir well and leave, covered, until a firm curd has formed. It should 'break' cleanly without leaving any ragged edges, otherwise it is not ready. Cut the curd into small cubes and leave to stand for 5 minutes. After this stir the curds round the whey by hand, but do not heat while this is going on. Continue stirring for half an hour, then leave to settle for ten minutes. Drain off the whey and ladle the curds into a clean cloth but do not fasten the cloth. An ideal way of arranging this is to put the curds into a large colander lined with a cloth and leave exposed to the air for fifteen minutes until a single mass of curd has been formed. Cut into blocks 15 cm (6 in) square but do not stack them on top of each other. Turn two or three times during the next fifteen minutes then break up the curds into nutmeg sized pieces and add salt at the rate of 28 gm salt to 1.3 kg curd (1 oz to 3 lbs) (2 tablespoons to 8 cups). Pack loosely into small moulds on mats, sprinkling a little chopped parsley between each layer. After two hours turn the cheeses upside down and continue doing this several times a day for the next two days. Remove them from the moulds and place on clean muslin on top of mats so that they continue to dry. After another two days rub a little butter on the outside and pack in cellophane paper or foil pricked with holes to let in air.

Munajuusto cheese
I am indebted to Mrs Helga Stewart who gave me the recipe that her mother used to make in her native Finland.

5 litres (1 gallon) (20 cups) fresh milk
½ litre (1 pint) (2½ cups) cultured buttermilk
2 eggs
1 teaspoon salt
1 teaspoon sugar

Whisk the eggs and then add to the buttermilk, stirring well. Add the mixture to the milk and heat very slowly, until the curds form and the mixture begins to thicken. Stop heating immediately and leave the curd, covered, in a warm place for an hour. Ladle out the curds into a cloth to drain.

The following day remove from the cloth, sprinkle on the salt, then put in a clean cloth with a board and weight on top to exert light pressure. Leave for three hours in this way then remove from the cloth. Sprinkle on the sugar and form the curd into a flattish, round 'cake'. Grill for a few minutes on each side until it browns slightly and it is ready for eating.

Mysost cheese
5 litres (1 gallon) (20 cups) fresh milk
5 litres (1 gallon) (20 cups) whey
½ litre (1 pint) (2½ cups) cream (optional)

This is a Norwegian whey cheese which is notable for its dark colour and sweet taste. Take equal quantities of milk and whey and cream if desired. (The cream will give a lighter colour and smoother texture.) Heat carefully so that it simmers without boiling over and stir frequently. Continue heating until it thickens and becomes fudge-like. Ladle onto boards and shape with Scotch hands. Leave to dry and wrap in foil until used. The sweet taste comes from the high sugar content of the whey, and the resulting cheese is nearly 40% sugar.

Neufchâtel cheese
5 litres (1 gallon) (20 cups) full cream milk
½ litre (1 pint) (2½ cups) cream (optional)
1 teaspoon commercial starter or 1 table-
* spoon home-made starter*
¼ teaspoon rennet

The best Neufchâtel cheese is made with very creamy milk, and that from the Channel Islands breeds is particularly suitable. If goats' milk is used, extra cream will be needed in the proportion of ½ litre to 5 litres (1 pint to 1 gallon) milk.

Use fresh milk and cream and heat to 21°C (70°F). Add starter and rennet and leave for 10 hours. Cut the curd into cubes and leave for 8 hours. Drain and suspend in a cloth. Add salt to taste then knead thoroughly. Shape into blocks with Scotch hands and wrap in foil until used. Different flavourings can be added as required during the kneading process. Some examples are chopped and crushed garlic cloves, chopped spring onions, ground black peppers or finely shredded capsicums (sweet peppers).

Parmesan cheese
5 litres (1 gallon) (20 cups) skimmed milk
2 teaspoons (2½ teaspoons) rennet

This is a hard Italian cheese for grating and is an essential ingredient in Spaghetti Bolognese. It takes a long time to ripen and dry properly. The minimum storage time is twelve months but it sometimes takes several years for this cheese to ripen properly, so perhaps you ought to start now.

Leave the evening's milk overnight and skim off the cream in the morning. Use the cream for something else. Heat the milk to 24°C (75°F) and add the rennet diluted in six teaspoons water so that curdling takes place within an hour. Raise the temperature to 55°C (130°F) and stir continuously, thereby breaking up the curd into small particles. Remove from the heat and leave undisturbed for 15 minutes. Drain off the whey and suspend the curd in cloths to finish draining. Next day, ladle the curd into lined moulds and after three days sprinkle salt on the surface. Next day remove from the moulds and sprinkle the other side. Leave to dry and cure for several months. If any moulds form on the surface, scrape them off and rub in a little salt but do not be tempted to rub the cheese with oil as some books recommend for this will impede the drying process. When it is ripened, grate the cheese and store in a jar with a close-fitting lid.

Pecorina Sardo – Ewe's milk cheese
I am indebted to Henryetta Edwards for this recipe.

10 litres (2 gallons) (40 cups) ewe's milk
1 teaspoon rennet

A friend in Sardinia witnessed and shared in the making of a Pecorina Sardo. The process began at dawn, when they had finished, a mature cheese was sliced and toasted on an open fire, and consumed with quantities of potent local wine – all before nine o'clock in the morning – but luckily not before he had written down the recipe for us:

Heat the milk in a suitable deep container, at a minimum temperature of 35°C (95°F). Add 'giagu' (rennet with herbs) in liquid form or ordinary rennet. Leave for half an hour, undisturbed by draughts or vibration. Stir gently to break up the curd, then knead, first with vertical pressure on the top surface, in handfuls, to finish up with a sausage-shaped 'loaf'. Cut the 'loaf' into manageable lumps. Remove each lump separately and squeeze out the surplus liquid before putting in a mould. Roughen the surface by pinching before adding the next. When the mould is full, press down to further expel the liquid. Turn the cheese over in the mould and put a weight on the top. Leave for an hour. Remove the cheese from the mould and immerse in fresh water. Then leave in salt water for 24 hours (caution – if water is too salty, so will the cheese be). Remove from the water, dry and smoke over an open fire of well-seasoned hard wood.

It is worth adding that safer results are obtained if a small commercial smoker is used for the 'smoking' process.

Pont L'Éveque cheese
5 litres (1 gallon) (20 cups) milk
1 teaspoon commercial starter of ½ cup (1 tablespoon) home-made starter
¼ teaspoon rennet
salt

This originated in the French town of that name and has only been made in Britain in limited quantities. The traditional way of making it was as follows:

Take a gallon of milk and add the starter then leave for half an hour. Heat to 30°C (86°F). Add rennet, then cover and leave for 50 minutes. Cut the curd into cubes, then cut diagonally 10 minutes later. Ladle the curds into cloths and suspend to drain. Tighten the cloth every 15 minutes for 4 hours, then ladle the curd into lined moulds. Place on mats to drain and turn on to fresh mats, three times during the next hour. Next day rub in salt and leave to stand on mats at a temperature of 18°C (64°F). When the white mould appears, remove to a cooler room until used.

It is worth adding that the best way of ensuring that you get the right mould is to use a commercial spray. Making a starter from shop-bought cheese is also a good way of ensuring that you achieve the right flavour and aroma.

Pot cheese
1 litre (2 pints) (5 cups) raw milk

This is the easiest cheese of all to make but it must be made from soured raw milk not pasteurized milk which has become sour. Leave the milk in a warm place, such as an airing cupboard, for 24 hours. When soured the following day, put in a saucepan and heat just enough to separate the solids from the whey, then drain off. Hang the curds in muslin to drain for a few hours, then work in salt and a little butter to taste. Put it in the refrigerator to firm and it is ready for eating straight away.

I dislike having things hanging up, dripping, and always avoid it unless I have to. For this type of cheese, and indeed for many other soft cheeses, I have found the Colwick cheese mould particularly suitable. It has small feet which keep it clear of the ground, and it will stand tidily on a cheese tray while the curd is draining.

1 Heat soured milk to separate curds and whey

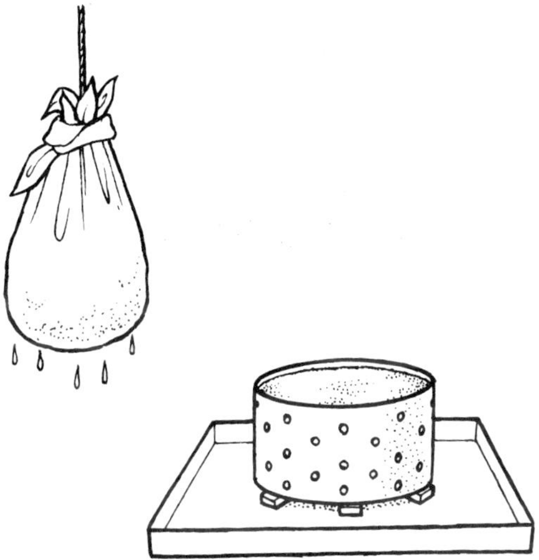

2 Drain either in muslin or in moulds on a tray

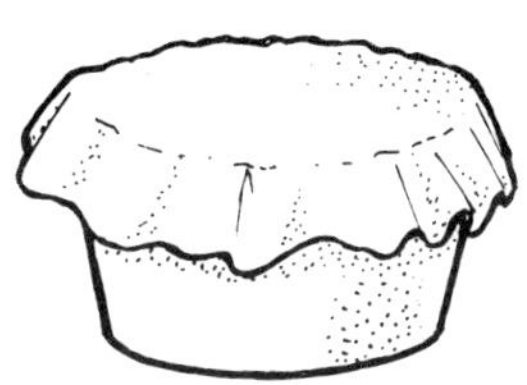

3 Keep covered while draining

Fig. 28 *Making a pot cheese*

Potato cheese

This was once widely made and eaten in Germany, and is a good way of utilizing left-over mashed potato and sour raw milk, or buttermilk.

Drain the sour milk or buttermilk through a cloth and then blend the remaining curd with mashed potato in the following proportions; 2 parts curd to 3 parts potato.

Add salt to taste and leave to ferment for one week, when it is ready for use.

Roman cheese

5 litres (1 gallon) (20 cups) previous night's milk
5 litres (1 gallon) (20 cups) morning's milk
½ cup fresh buttermilk as starter (necessary only if all-fresh milk is used)
1 teaspoon rennet

Roman or Romano cheeses originated in Italy and are used as table cheeses after a short period of ripening or as hard grating cheeses when ripened for six months.

Take a mixture of evening and morning's milk and heat to 49°C (120°F). Cool immediately to 32°C (90°F) and add the starter. Leave for an hour, then add rennet. Leave until the curd is hard then cut into cubes and heat for 45 minutes, gradually increasing the temperature to 38°C (100°F) for the last 15 minutes. Stir gently with the hands during this time, then remove from the heat and leave undisturbed for an hour. Drain off the whey, cut the curd again and leave to drain in suspended cloths for two hours. Pile into moulds lined with cloths and press for 24 hours. Upturn and press for another 24 hours. Remove and immerse in brine for three hours, then rub the surface with dry salt and bandage. Leave for a week in a cool place, turning daily, then remove bandages and wipe with a clean cloth. Dip in hot water and seal off any holes or cracks. Cover in wax after one week and store for six months if required as a grating cheese similar to Parmesan. Alternatively, it can be eaten after three months. If this recipe is made with cows' milk the cheese is called Vacchino Romano, while that made from ewes' milk is Pecorino Romano. Where goats' milk is used the cheese is called Caprino Romano.

Roquefort cheese

5 litres (1 gallon) (20 cups) ewes' milk or Jersey milk
1/2 litre (1 pint) (2 1/2 cups) cream if goats' milk or thin cows' milk is used
1 teaspoon commercial starter or 1 dessertspoon home-made starter
1/4 teaspoon rennet

This famous French cheese developed as a result of the unique conditions in that region; the sheep with their rich milk, and the presence of the Roquefort caves. These caves seldom have a temperature that exceeds 10°C (50°F) and the degree of humidity is at an optimum level for curing cheese.

The traditional way of making it is as follows: Heat ewes' milk to 27°C (81°F) and add rennet. Leave to coagulate for two hours, then cut into cubes. Stir gently until the curd portions are pea-sized, then drain off the whey by hanging the curds in a cloth overnight. Traditionally, in order to innoculate the curds with the famous mould, a fresh loaf was taken from the oven and dipped in paraffin. A pipette was then used to introduce spores of the mould into the centre of the loaf which was then left for a month in a dark moist place. The bread was then ground up and mixed with the curd in the proportion of 1 part innoculated bread to 10,000 parts of curd. The curd was then put in moulds and cured in the caves. For the home cheesemaker, a commercial sachet of *Penicillium roqueforti* can be introduced (see page 60) or small shreds of the mould from a piece of shop-bought Roquefort can be mixed in with the curd. Sprinkle on a little salt, put the curds in moulds on mats and leave to settle naturally at a temperature of 10°C (50°F).

Turn the moulds every day for the next four days. Remove from the moulds and rub with salt. Pierce the cheese with a stainless steel needle and leave to ripen for three weeks. After this time rub a little vegetable oil on the outside to stop it drying up. Wrap in cellophane or greaseproof paper.

Sage cheese

This is Cheddar cheese which has had a mixture of sage and spinach leaves added at the rennetting stage.

Wash six leaves of sage and three of spinach then put them in a blender with a little hot water. If a blender is not available, pound them with a pestle and mortar and force through a muslin bag. Add with the rennet then follow the instructions given for Cheddar cheese on page 69. This mixture of sage and spinach can also be used with other cheeses such as Derby and Double Gloucester.

Skimmed milk cottage cheese

5 litres (1 gallon) (20 cups) skimmed milk
1 teaspoon starter
1/2 teaspoon salt

Heat the skimmed milk to 24°C (75°F) and add a small amount of starter to speed up curdling. After a day, cut into pieces and place in a basin over a saucepan of hot water. Leave for 30 minutes at 30°C (86°F) stirring occasionally. Place the curd in a cloth and hang it up to drain. Add the salt and work in well.

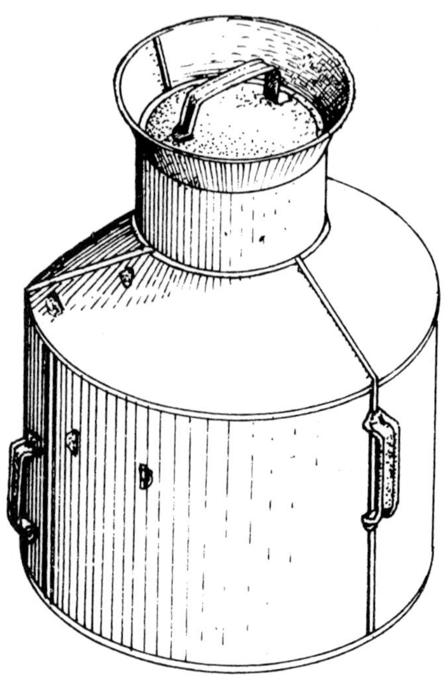

Smallholder cheese

I am indebted to Vicky Hartley for this recipe.

15 litres (3 gallons) (60 cups) milk
125 ml (4 fl oz) (½ cup) starter
5 ml (1 teaspoon) rennet
28 g (1 oz) (2 tablespoons) salt.

This is a semi-hard cheese but quite firm and mild. This quantity will produce approximately 1.3 kg (3 lbs) of cheese. I had not come across this recipe until I attended an excellent cheesemaking course which was run by Vicky Hartley at Quainton Dairy in 1976, which is where I first made it. The basic steps are the same as those for a Cheddar.

Heat the milk to 68°C (155°F) then cool immediately to 32°C (90°F). Add the starter, stir and leave for half an hour. Mix the rennet with four teaspoonfuls of previously boiled and cooled water and stir into the milk. Top-stir the milk to mix in the cream, then leave for 30–40 minutes until the curd is firm and does not leave a milk stain on the back of the finger. Cut the curd as shown on page 56 and leave until whey shows at the top. Increase the heat to 38°C (100°F) over a period of half an hour, stirring the curds by hand during this time. Stop heating and let the curds settle for another half an hour. Drain the curds into a cloth and tie up into a bundle. Open the cloth after 15 minutes and cut the solid curd into four slices. Stack them on top of each other in the way that was described for Cheddar (p. 69), and alternate the positions of the slices every 15 minutes by putting the outside ones in the middle and vice versa. This process is known as 'cheddaring'. Break the curd into pieces the size of a nutmeg and sprinkle on the salt. Line the cheese mould with a muslin cloth which has been boiled then press the curd firmly into the mould. Put in the cheese press under a light pressure and gradually increase it. Next day remove the cheese from the mould and put it back upside down, increasing the pressure. The following day take the cheese out of the mould and if there are no cracks in it dip it in water at 66°C (150°F) for 30 seconds. Return to the press and leave for another five days, turning once a day if possible. Take it out of the mould and leave to cool and dry. It can either be bandaged or waxed.

Soda skim cheese

I am indebted to an American friend, Hilda Short, for this recipe. It is described in her own words.

15 litres (3 gallons) (60 cups) skimmed milk
2 teaspoons (2½ teaspoons) rennet
1 teaspoon bicarbonate of soda
85 gm (3 oz) (6 tablespoons) butter
500 ml (8 fl oz) (1 cup) sour cream
2 teaspoons (2½ teaspoons) salt

This may not sound very appetizing, but it's what I call it for convenience, and it's a good cheese. Take 15 litres (3 gallons) of skimmed milk and add 2 teaspoonfuls of rennet. Leave it until it has clabbered (coagulated) then heat slowly until it reaches 35°C (95°F). Maintain at this temperature for half an hour, then drain the liquid away. Put the curds in a cloth and press it with the hands to squeeze as much whey out as possible. Hang it for a while. Mix the bicarbonate of soda and butter into the curds, then leave to stand for 2 hours. Add a cup of sour cream and two teaspoonfuls of salt and heat slowly to blend the mixture and get rid of any lumps. Ladle into lined moulds to drain. The cheese is ready to eat in 5 days.

Tuscany sheep's milk cheeses

For this recipe I am indebted to Susan Wrightson of Tuscany whose description is given below.

50 litres (11 gallons) (9 gallons) sheep's milk
1 tablespoon (1½ tablespoons) rennet

This recipe makes two different cheeses – a hard cheese and a Ricotta. Sheep's milk is rich, creamy and slightly sweeter than cows'. You can make cheese with as little as half a gallon, but it is much more convenient to make a decent quantity at a time. When our friendly neighbourhood shepherd, here in Tuscany, grazes his flock in the vicinity, we take our spare sheep's milk up to him twice a day until we have given him 50 litres (11 gallons). He then gives us a 50-litre churn full of fresh milk from one milking (one makes this cheese from fresh milk) and we make our cheese.

You need a huge, spotlessly clean pan, a small wood fire – preferably last year's oak, and an iron pan support to fit over it. You also need cheese moulds with perforations for draining the whey, rennet, a wooden spoon, ladle and a split bamboo for cutting the curd. When the fire is at the red ash stage, empty all except 1 litre of the fresh milk into the pan and put the spare litre in a clean jug for later use. Heat the milk until hand-hot then remove from the fire; you will need two people to lift it. Stir in a tablespoonful of rennet and leave for 15–20 minutes, or until the curd has set.

With your hands under the surface gently break up the curd, then leave for a moment or two. Now take a cheese mould and gently press it base downwards through the whey to send the curd down. Try to get it all down so that the second cheese – the Ricotta or 'recooked' cheese will be smooth. Leave for a few minutes then take your split bamboo stick (or knife) and cut the curd, two cuts each way at right angles to each other. Then place two flat sticks across the pan as a support for the cheese mould. Gently pick up handfuls of curd and press down into the mould. Continue until the

mould is full and domed on top. Empty the pressed curd from the mould into one hand (it should all adhere nicely in its shape), change into the other hand and replace it upside down in the mould. Press down well and repeat the turnings three times with each mould. Leave to drain and covered with muslin. While you are waiting for the cheese to dry you can make the Ricotta. Return the pan of whey to the fire and add the litre of fresh milk that was set aside earlier. Watch the whey carefully, stirring all the time. Just before the liquid boils, the Ricotta will rise – magical, creamy clouds of it – to the surface. Stop stirring as soon as this happens and when the liquid boils remove the pan from the heat. Take out the Ricotta with a perforated ladle and fill the remaining cheese moulds, without pressing down, but allow to drain well. This Ricotta will not keep any longer than fresh milk but is delicious eaten with salt and herbs, or with sugar and fruit such as strawberries. We usually keep one for ourselves and take the others up to the village to sell. Having made your Ricotta, return now to the true cheeses. Put each cheese mould with contents, into the hot whey, the heat forming the rind. Lift out, press, turn and immerse again, and finally lift out to drain. When quite cold, lightly cover the tops with salt. After 24 hours take the cheeses out of the moulds and salt the remaining surfaces. After another 24 hours, wash the cheeses gently in cold water. Put them to mature on a well-scrubbed, lightly salted wooden board and turn them every day for a month. At this stage rub them with a light film of sunflower oil. After this they must be turned every few days.

You can start eating the cheese at a month old, and uncut ones will keep at least four months. It is a good idea to rub at least one with wood soot from the smoke of well-seasoned hardwood, to form Pecorina Grano – a hard grating cheese like Parmesan. These recipes will work just as well for goat's milk, but there will be less cheese as the milk is not as rich.

Wensleydale cheese

*5 litres (1 gallon) (20 cups) previous night's
 milk*
5 litres (1 gallon) (20 cups) morning's milk
*½ cup (1 tablespoon) home-made starter or 1
 teaspoon commercial starter (necessary only
 if all-fresh milk is used)*
½ teaspoon rennet

This is the famous Yorkshire cheese. Leave
the evening's milking overnight and skim
the cream off in the morning. Add the
morning's milk to the skimmed and warm
the cream, before adding that too. Heat to
30°C (86°F) and add starter. When the acidity is at 0.19% after about half an hour, add
the rennet and stir well for five minutes.
After an hour, the curd should be firm and
ready for cutting. Cut into cubes and start
to stir gently with the hands, for the curd is
softer than for many other cheeses. Gradually raise the temperature to 31°C (88°F)
and continue stirring. Stop heating and
leave to stand until the acidity of the whey
is between 0.14–0.16% (about 30 minutes).
Drain off the whey and put the curd into
draining cloths on a board. Pour on some of
the whey again and leave until an acidity of
0.2% is reached after about 15 minutes.
Drain and cut the curd into walnut sized
pieces, and add salt in the proportion of 14
gm to 1.3 kg curd (½ oz to 3 lbs) (1 tablespoon to 8 cups). Take care not to break up
the curd too much while salting is taking
place. Ladle into lined moulds and leave
overnight. In the morning, put into fresh
moulds and apply pressure gradually
through the day. The next day, turn and
apply pressure again. In 24 hours, remove
from the press, dry, bandage and leave to
ripen for a month. Dry white wine or dry
cider is nice with this cheese.

Yoghurt Liptauer Cheese

For this recipe I am indebted to Angela Lyle
of Ivybridge in Devon.

1.5 litres (3 pints) (7½ cups) milk
1 carton fresh yoghurt
225 gm (½ lb) (1 cup) butter
50 gm (2 oz) (⅜ cup) anchovy fillets
*1 tablespoon (1½ tablespoon) chopped gher-
 kins*
*1 teaspoon each of coriander, dry mustard,
 paprika*
Pinch of herb salt.

Warm milk to between 41°C–45°C (105°F–
115°F). Stir in the carton of fresh yoghurt.
Put in a heated thermos flask for 4–8 hours.
The thermos is heated and sterilized by
washing out with boiling water. Take off
the lid and keep in the refrigerator until
ready for use. Now drain off the yoghurt in
muslin in the sink overnight. Beat the butter and add to the yoghurt curds. Beat in
the anchovy fillets, gherkins, coriander, dry
mustard, paprika and herb salt, then refrigerate until needed. It is ready for eating
straight away but will last for up to a week
in the refrigerator.

Recipes using cheese

Cheese spread

*225 gm (½ lb) (1½ cups) soft cheese or grated
 hard cheese*
50 gm (2 oz) (4 tablespoons) butter
4 tablespoons (6 tablespoons) milk
½ teaspoon salt
½ teaspoon mustard
1 teaspoon Worcestershire Sauce

Cream the butter and add the cheese and
seasonings. Add the milk and beat well to
acquire a smooth consistency. Put into a
dish or plastic container and store in the
refrigerator until use. It will keep for about
four days. Different flavourings can be
added to vary the basic recipe as follows:
chopped spring onions or chives, parsley,
tomato puree, chopped celery, anchovies,
mustard and cress or watercress, chopped
radishes, nuts, crushed garlic.

Welsh Rarebit
225 gm (¹/₂ lb) (1 cup) grated Caerphilly or
  Cheshire cheese
250 ml (¹/₂ pint) (1¹/₄ cups) milk
25 gm (1 oz) (1¹/₂ tablespoons) butter
2 tablespoons (3 tablespoons) flour
Pinch of salt
Pinch of paprika

Add the flour to the butter and heat gently until a 'roux' is formed. Add the milk, cheese and seasoning and stir well. Heat gently stirring all the time, until the mixture thickens. Pour over buttered toast and, just before serving, sprinkle on the paprika to decorate.

Cheese Fondue
225 gm (¹/₂ lb) (1 cup) grated Cheshire, Caer-
  philly or Cheddar cheese
1 garlic clove
25 gm (1 oz) (1¹/₂ tablespoons) butter
4 tablespoons (6 tablespoons) white wine
4 tablespoons (6 tablespoons) milk
1 dessertspoon flour
Pinch of black pepper

Crush the garlic clove and add to the wine. Leave to stand and meanwhile heat the butter and flour until mixed. Add the cheese, wine and milk and heat slowly, stirring all the time until the mixture thickens. Sprinkle on the black pepper and serve with cubes or fingers of bread.

Welsh cheese pudding
225 gm (¹/₂ lb) (1 cup) grated Caerphilly,
  Cheshire or Cheddar cheese
250 ml (¹/₂ pint) (1¹/₄ cups) milk
110 gm (4 oz) (4 cups) breadcrumbs
2 eggs
1 teaspoon mustard
¹/₂ teaspoon salt
Pinch of black pepper

Separate the eggs and whip the whites until stiff. Meanwhile heat the milk and stir in the breadcrumbs. Remove from the heat and stir in the egg yolks, cheese and season-ings. Leave to cool for ten minutes then fold in the egg white. Place in a greased dish, sprinkle a little extra grated cheese on the surface and bake in an oven at 200°C (400°F) or Gas Mark 6 for half an hour, until the top is nicely browned.

Cheesecakes

Once you start making soft cheese you tend to get carried away and your family may well start being rude about the never-ending supply. If that happens, do not take any notice – give them cheesecake.

Cheesecakes are delicious but weight-watchers are advised to exercise mod-eration. They are made in two stages, the base and the filling. With some recipes there may be a third stage, if there is a separate topping. The base is usually pastry baked 'blind', with dried peas in it to keep the shape and then discarded after cooking. An alternative and quick method is to use crumbled digestive biscuits.

Pastry base
160 gm (6 oz) (scant 3 cups) plain flour
75 gm (3 oz) (6 tablespoons) butter or mar-
  garine
Pinch of salt
Water

Sift the salt into the flour and rub in the fat, adding just enough water to bind. Roll out on a floured board and line an oiled 30 cm (12 in) plate. Fill with dried peas and bake in a hot oven at 220°C (425°F) or Gas Mark 7 for 10–12 minutes. The dried peas keep the shape of the base, ensuring that the sides do not crumble and that the bottom does not dome. They can be stored in an airtight container after use and kept for the same purpose on a future occasion.

Biscuit base
*1 packet plain digestive biscuits (225 gm) (¹/₂
  lb)*
140 gm (5 oz) (10 tablespoons) butter
¹/₄ teaspoon ground nutmeg
¹/₄ teaspoon ground cinnamon

Crush the biscuits with a rolling pin and
sprinkle on the cinnamon and nutmeg. Melt
the butter in a saucepan and add the crumbly mixture. Mix well and press into an
oiled cake tin. Leave in the refrigerator
until quite cold. The biscuits give the base a
much crunchier texture and, in my view,
this base is more interesting than the pastry base.

Lemon cheesecake
125 ml (¹/₄ pint) (⁵/₈ cup) milk
2 eggs
1 sachet gelatine
2 tablespoons (3 tablespoons) sugar
2 lemons
225 gm (¹/₂ lb) (1¹/₄ cup) soft cheese

Separate the eggs and beat the yolks with
the milk. Add the gelatine and sugar
together with the grated rinds of the
lemons. Heat the mixture gently until the
gelatine dissolves, then leave to cool. Beat
the soft cheese until it is smooth then whisk
the egg whites separately until stiff. When
the original mixture is quite cold add the
lemon juice from the two lemons and stir in
the cheese. Fold in the egg whites and pour
the mixture onto the base. Leave in the
refrigerator to set.

Orange cheesecake
285 gm (10 oz) (1¹/₂ cups) soft cheese
¹/₂ packet orange jelly
3 tablespoons water
1 tablespoon sugar
1 orange

Dissolve the jelly in water and heat carefully until dissolved. Stir in the sugar and the
grated rind and juice of the orange. Beat
the cheese until smooth and stir in the
orange mixture. Add to the base and leave
to set in the refrigerator. Before serving,
segments of tinned oranges can be placed
on top if desired.

USEFUL ADDRESSES

United Kingdom

Organizations and Associations

Ministry of Agriculture, Fisheries & Food (MAFF), Whitehall Place, London, SW1A 2HH.

Agricultural Development and Advisory Service (ADAS), Great Westminster House, Horseferry Road, London, SW1P 2AL.

Rare Breeds Survival Trust, 4th Street, National Agricultural Centre, Stoneleigh, Kenilworth, CV8 2LG Warks.

The National Institute for Research into Dairying, Shinfield, Reading, RG2 9AT, Berks.

The Hannah Research Institute, Ayr, KA6 5HL, Scotland.

The Agricultural Research Institute of Northern Ireland, Hillsborough BT26 6DP, Co. Down, Northern Ireland.

Commonwealth Bureau of Dairy Science and Technology, Lane End House, Shinfield, Reading, RG2 9BB, Berks.

Royal Association of British Dairy Farmers, Robarts House, Rossmore Road, London, NW1 6NP.

The British Goat Society, Rougham, Bury St. Edmunds, Suffolk, IP30 9LJ. This is the national society which will provide details of local goat societies which are affiliated to it.

Goat Veterinary Society, Windrush Cottage, 84 Bradenstoke, Chippenham, Wilts.

The Jersey Cattle Society of Great Britain, 154 Castle Hill, Reading, Berks.

British Friesian Cattle Society of Great Britain and Ireland, Scotsbridge House, Rickmansworth, Herts. WD3 3BB.

English Guernsey Cattle Society, Giggs Hill Green, Thames Ditton, Surrey.

Ayrshire Cattle Society, PO Box 8, 1 Racecourse Road, Ayr, Scotland.

South Devon Herd Book Society, 24 Courtenay Park, Newton Abott, Devon.

Welsh Black Cattle Society, 13 Bangor Street, Caernarfon, Gwynedd LL55 1AP, Wales.

The Dexter Cattle Society, Lomond, Seckington Lane, Newton Regis, Tamworth, Staffs.

British Canadian Holstein Society, 69 Thoroughfare, Woodbridge, Suffolk, IP12 1AH.

The British Friesland Sheep Society, Russley Park, Baydon, Wilts. SN8 2JY.

Milk Marketing Board, Thames Ditton, Surrey.

National Dairy Council, 5 St. John, Princes Street, London, W1

Butter Information Council, Bank Street Suite, 158 High Street, Tonbridge, Kent TN9 1BJ.

The Cheese Bureau, 40 Berkeley Square, London W1X 6AD.

Farmhouse English Cheese Information Office, 16 Bolton Street, London, W1Y 8HX.

Suppliers

Self Sufficiency & Smallholding Supplies, The Old Palace, Wells, Somerset. Suppliers of a wide range of dairying and cheesemaking equipment and supplies.

R.J. Fullwood & Bland Ltd., Ellesmere, Salop. Milking equipment, cream separators, dairy supplies and commercial yoghurt cabinets.

Cockx & Sons (London) Ltd. Windham Road, Chilton Industrial Estate, Sudbury, Suffolk CO16 6XD. Yoghurt vats, and plastic yoghurt pots.

R. Chadwick & Son (Bury) Ltd. Villiers Street, Bury, Lancs. Hand-operated foil capping machines for yoghurt and cream pots.

Lakeland Plastics Ltd., 38 Alexandra Buildings, Windermere, Cumbria, LA23 1BQ. Plastic goats' milk bags, heat sealers and yoghurt pots.

Landkey-Newland, 7 Brethren Trading Estate, Barnstaple, N. Devon. Small milking machines.

R. & G. Wheeler, Hoppins, Dunchideock, Exeter, N. Devon. Cheese presses and cheese vats.

Gliddon & Squire, Tuly Street, Barnstaple, Devon. Small milking machines.

Halifax Scale Sales Ltd., Brighouse Road, Hipperholme, Halifax HX3 8EF. Milk recording scales.

D. Malpas Fabrications, 31 Regent Street, Barwell, Leicester, LE9 8GY. Goat milking stands and hayracks.

Astell's Laboratory Company, 172 Brownhills Road, Catford, London SE6 2DL. Lloyd's Acidmeters and Lewis bottles.

Chris Hansen Laboratories Ltd., 476 Basingstoke Road, Reading, RG2 0QL. Cheese and yoghurt freeze-dried cultures. Cheese rennet. Annatto. Mould spores for blue and white moulded cheeses (*Pencillium roqueforti* and *pencillium candidum*).

Anything Goats, Manor Cottage, Wretton Road, Stoke Ferry, King's Lynn, Norfolk PE33 9QJ. Goat and dairying supplies.

Lincolnshire Smallholders Supplies Ltd., Willow Farm, Thorpe Fendyke, Wainfleet, Lincs. PE24 4QH. Goat's milk bags and dairying supplies.

Robby Tanks, Cruwys Morchard, Tiverton, Devon. Small milk pasteurizing units.

Tefal Housewares Ltd., Rivermeade, Oxford Road, Uxbridge. Ice cream makers.

Hosier Farming Systems, Collingbourne Ducis, Marlborough, Wilts. SN8 3EH. Small milking machines.

F. Ritson, Goat Appliance Works, Longtown, Carlisle. Goatkeeping equipment.

Courses

Most dairying and cheesemaking courses are advertised in *Home Farm* magazine, Broad Leys Publishing Company, Widdington, Saffron Walden, Essex, CB11 3SP.

The following organizations and individuals run regular courses.

Mr & Mrs Rayner, Penllan, Bwlchyffridd, Newton, Powys, Wales.

Mrs May, Priestlands Goat Farm, Claygate, Marden, Kent.

Yarner Trust, Beacon Farm, Dartington, Devon.

Mrs C.J. Brown, Cherrytree House, Dunton Bassett, Leics.

It is also worth contacting the nearest agricultural college to see if an appropriate course is available.

Museums

The following museums have extensive dairying displays.

National Dairy Museum, Wellington Country Park, Nr. Stratford Saye, Reading, Berks.

Easton Farm Park, Wickham Market, Woodbridge, Suffolk.

Museum of East Anglian Life, Abbot's Hall, Stowmarket, Suffolk.

Museum of English Rural Life, University of Reading, Whiteknights, Reading, Berks.

Somerset Rural Life Museum, Abbey Barn, Glastonbury, Somerset.

Wantage Museum, Civic Hall, Portway, Wantage, Oxfordshire.

Countryside Museum, Bicton Gardens, East Budleigh, Devon.

Castle Museum, York.

United States of America

Organizations

United States Department of Agriculture (USDA), Washington DC 20505.
The American Dairy Goat Society, Box 186, Spindale, North Carolina 28160.
The American Jersey Cattle Club, 2105 J. South Hamilton Road, Columbus, Ohio 43227.
The Holstein-Friesian Association of America, Box 808, Brattleboro, Vermont 05301.

Suppliers

Cumberland General Store, RR3, Crossville, Tennessee 38555. General dairying equipment.
New England Cheese Making Supply Co., Box 85, Ashfield, Massachusetts, 01330.
American Supply House, PO Box 1114, Columbia, Montana 65201. Goat equipment
Chr. Hansen's Laboratory Inc. 9015 W. Maple Street, Milwaukee, Wisconsin, 53214. Cheese and yoghurt cultures, rennet, annatto and vegetarian rennet.
Marika Dairy Cultures, Natalex Corp. Box 3060, Weehawken, New Jersey 07087.
Caprine Supple Catalog, 6657 Woodland Road, Shawnee, KS 66218. Cheesemaking and goat equipment.
Countryside General Store, Highway 19 East, Waterloo WI 53594. Cheesemaking equipment.
Plow and Hearth, Box 530, Madison, VA 22727. Dairying and cheesemaking supplies.
Nasco, 901 Janesville Avenue, Fort Atkinson, WI 53538. Cheesemaking supplies.

Canada

Organizations

Ministry of Agriculture Canada, 930 Carling Avenue, Ottowa, Ontario.
Ontario Dairy Goat Society, RR1 Millgrove, Ontario, LOR 1VO.
Canadian Jersey Cattle Club, 343 Waterloo Avenue, Guelph, Ontario NIM 3KI.
The Canadian Cattle Breeders Society, Roxton Pond, Shefford, Quebec JOE 1ZO.
Holstein-Friesian Association of Canada, Brantford, Ontario.
Alberta Goat Breeders Association, RR 6, Site 6, Box 16, Edmonton, Alberta.
Alberta Goat Information Centre, Box 89, Kingman, Alberta TOB 2MO.

Suppliers

The Pioneer Place, Route 4, Aylmer, Ontario N5H 2RE. Cream separators.
Berry Hill Ltd., 75 Burwell Road, St. Thomas, Ontario N5P 3R5. Ice cream makers.
Contempra Industries Inc., 371 Essex Road, Tinton Falls, New Jersey 07753. Yoghurt makers.
Ketchum Manufacturing Sales Ltd., 396 Berkley Avenue, Ottowa. General livestock equipment.
Braun Natural Yoghurt Maker, 3269 American Drive, Mississauga, Ontario L4V 1B9.
Horan-Lally Co. Ltd., 1146 Aerowood Drive, Mississauga, Ontario L4W 1Y5. Cheese and yoghurt making supplies.

Australia

Organizations

Department of Agriculture (DAF), 25 Grenfall Street, Adelaide 5000.
The Goat Breeders Society of Australia, Box 4317, GPO Sydney, NSW 2001.

Suppliers

Self Sufficiency Supplies, 256 Darby Street, Newcastle 2300. Dairying equipment.
Self Sufficiency & Smallholding Supplies, The Old Palace, Wells, Somerset, England. This company will ship orders direct to Australia and New Zealand.
Victorian Rennet Manufacturing Co. Pty., 16 Queen Street, Munawaging, Victoria 3131. Rennet.

Department of Agriculture & Fisheries, Bag,
  1671, GPO Adelaide 5001. Yoghurt
  cultures.

New Zealand

Organizations
Ministry of Agriculture, Private Bag,
  Wellington.
New Zealand Association of Small Farmers,
  PO Box 2081, Palmerston.
Goat Breeders Association, PO Box 294,
  Morrinsville.

BIBLIOGRAPHY

Periodicals

Home Farm (formerly Practical Self
  Sufficiency), Broad Leys Publishing Co.,
  Widdington, Saffron Walden, Essex. CB11
  3SP (UK).
Mother Earth News, Box 70, Hendersonville,
  NC 28739 (USA).
Countryside, 312 Portland Road, Waterloo,
  Wisconsin 53594 (USA).
Hoards Dairyman, Fort Atkinson, WI 53538
  (USA).
United Caprine News, Box 27, Leakey, Texas
  78873 (USA).
Dairy Goat Journal, Box 1808, Scottsdale,
  Arizona 85252 (USA).
Acres USA, 10227, East Sixty-First Street,
  Raytown, MO 64133 (USA).
Cheesemakers' Journal, Box 85, Ashfield, MA
  01330 (USA).
Harrowsmith Magazine, Queen Victoria
  Road, Camden East, Ontario KOK 1JO
  (Canada).
Grass Roots, Box 900, Shepparton 3630
  (Australia).

Earth Garden, PO Box 378, Epping, NSW
  2121 (Australia).
The Small Farmer, PO Box 2081,
  Palmerston, North Island (New Zealand)

Books on dairying and animal husbandry

Bode, A. and Brooks, H., *Australian Goat
  Husbandry*.
Castle, M.E. and Watkins, P. *Modern Milk
  Production*, Faber (UK), 1979.
Downing, E., *Keeping Goats*, Pelham,
  Books (UK), 1976.
Hetherington, L., *Home GoatKeeping*, E.P.
  Publishing (UK) 1977.
Hetherington, L., *All About Goats*, Farming
  Press (UK), 1977.
Jeffrey, H.E., *Goats*, Cassell (UK), 1975.
Loon, D. van, *The Family Cow*, Garden Way
  (USA), 1976.
Luisi, B., *A Practical Guide to Small Scale
  Goatkeeping*, Rodale (USA), 1979.

Mackenzie, D., *Goat Husbandry*, Faber (UK), 1980.

Mills, O., *Practical Sheep Dairying*, Thorsons (UK), 1982.

Minter, P.V., and Rogers, F., *Goats. Their Care and Breeding*, KR Books (UK) 1979.

Peck, P.C., *Your own Dairy Cow*, Thorsons (UK), 1979.

Russell, K., *The Principles of Dairy Farming*, Farming Press (UK).

Salmon, J., *The Goatkeeper's Guide*, David & Charles (UK), 1976.

Shields, J., *Exhibition and Practical Goat Keeping*, Spur Publications (UK), 1977.

Spreckley, V., *Keeping a Cow*, David & Charles (UK).

Staniland, I., *Goatkeeping*, Thorsons (UK), 1979.

Thear. K., *The Family Smallholding*, B.T. Batsford Ltd., (UK), 1983.

British Goat Society booklets:
Goat Feeding
Breeds of Goats
Dairy Work for Goatkeepers
Goatkeeping
Life Story of a Goat
Wild Food for Goats.

Practical Goat-keeping. John & Jill Halliday, Ward Lock (UK) 1982.

Goat Production in the Tropics. C. Devendra and M. Burns. Commonwealth Agricultural Bureaux (UK) 1970.

MAFF booklets (free on application from MAFF Publications, Willowburn Trading Estate, Alnwick, Northumberland NE66 2PS):
Dairy Goatkeeping No. AL 18
Cream No. AL 495
Clotted Cream No. AL 438
Farmhouse Production of Cream Cheese No. AL 222.
Taints in Milk. No. AL 458.
Starters for Cheesemaking. No. AL 302.
Hand Cleaning of Dairy Equipment. No. AL 422.

Books on cheese and other dairy produce

Axler, B., *The Cheese Handbook*, Cassell (UK), 1970.

Black, M., *100 Ways with Cheese*, Letts (UK), 1976.

Black, M., *Home-made Butter, Cheese and Yoghurt*, E.P. Publishing (UK), 1977.

Carroll, R. & R., *Cheesemaking Made Easy*, Garden Way (USA), 1982.

Cheke, V., and Sheppard, A., *Butter and Cheese Making*, Prism Press (UK), 1980.

Cooper, M., *Discovering Farmhouse Cheese*, Shire Publications (UK), 1978.

Glynn, C., *Cheese and Cheese Making*, Macdonald (UK), 1977.

Hunt, C.J., *Home Cheese Making*, Country Things (USA), 1974.

Lewis, B.R., *Farmhouse English Cheese*, E.P. Publishing (UK), 1977.

Newman, R., *The Sainsbury Book of Cheese*, Cathay Books (UK), 1982.

Nilson, B., *Cooking with Yoghurt and Cheese*, Granada (UK), 1979.

Nilsson, A., *The Art of Home Cheesemaking*, Woodbridge Press (USA), 1979.

Ogilvy, S., *Making Cheeses*, B.T. Batsford Ltd. (UK), 1977.

Orga, I., *Cooking with Yoghurt*, Coronet (UK), 1981.

Pike, M.A., *Soft Cheese Craft*, Whittet Books (UK), 1982.

Acidity testing 48–9
Acidmeter 40, 48–9
Annatto 45, 62
Appetitost cheese 66
Apple yoghurt mousse 32

Bacteria 19, 26, 28, 40, 49
Bandaging cheese 59
Bleu des Causses 47
Bondon cheese 66
Brandy butter 46
Brie cheese 67
Butter, churning 43–4; colouring 45; icing 46; moulds 40; muslin 53; recipes 46; salting 44; washing 44; working 44–5;
Buttermaking, equipment 40–1; problems 45–6
Buttermilk 31, 44; cheese 67
Butterworker 45

Caerphilly cheese 67–8
Cambridge cheese 68
Cantal cheese 68
Caramel ice cream 38–9
Cattle 9
Caws bach cheese 69
Channel Islands milk 22
Cheddar cheese 69–71
Cheddaring 55
Cheese, cloth 53; colouring 62; fondue 88; mats and trays 53; moulds 53; press 53; selling 65; spread 87; types 47–8
Cheesecakes 88–9
Cheesemaking, advice 52–3; equipment 53–5; pests and problems 62–5; process 55–60; records 62
Cheshire cheese 71
Chocolate ice cream 37
Churning butter 43–4
Club cheese 71
Coffee ice cream 38
Colby cheese 71–2
Cole slaw 32
Colouring, butter 45; cheese 62
Colwick cheese 72

Condensed milk 22
Cornflour custard ice cream 38
Cottage cheese 84–5
Coulommier cheese 72
Cream, cheese 74; clotted 35–6; pasteurizing 35; selling 39; separation 33–5; sour 36; storage 35
Crowdie cheese 74
Curd cheese 79; knife 53, 55
Curds 55–8
Custard, home made 24

Dairy 17–20; animals 9–13; thermometer 40
Derby cheese 74–5
Devon farmhouse cheese 75
Dorset blue cheese 75
Double Gloucester cheese 76
Double saucepan 53
Dried milk 22
Dunlop cheese 76

Edam cheese 76–7
Evaporated milk 22
Ewes' milk 26–7, 47, 77, 82, 84, 86; cheese 82

Feta cheese 77
Fore-milk 15
Freezing, butter 45; cream 35; milk 23
French goats' cheese 77

Gervais cheese 78
Ginger ice cream 38
Goats 10–11; cheese 77, 79; milk 22, 23, 24, 26, 28–9, 41, 66
Gorgonzola cheese 78
Gouda cheese 78–9
Gruyère cheese 79

Herb butter 46
Homogenized milk 22

Ice cream, making 36–7; recipes 37–9
Incubating starter culture 52
Innoculating milk 51

Jamaican ice cream 38
Junket 25

Keeping cheese records 62
Kefir 30–1
Koumiss 31

Lactic curd goats' cheese 79
Lactic ferment culture 53
Ladies' bedstraw 48
Leicester cheese 79–80
Lemon cottage cheese 80
Lemon ice cream 39
Lewis bottles 52
Little Welsh cheese 80

Maple syrup and walnut ice cream 38
Mastitis 12–13
Menyn bach 43–4
Milk, constituents 21; recipes 24–5; recording 16; selling 24; sheep 11–12; storing 23; treatment 23
Milk types, Channel Islands 22; condensed 22; dried 22; evaporated 22; homogenized 22, 43; pasteurized 22; sterilized 22; untreated or raw 22
Milk fat content 21, 28–9, 33
Milking 13–16
Milling curds 55–8
Moulding cheese 58
Moulds, blue 60; white 60–2

Neufchâtel cheese 81

Oiling cheese 59–60

Parmesan cheese 81
Pasteurized milk 22
Pasteurizing, milk 23; cream 35, 40
Pecorina Sardo 82
Penicillium, camemberti 62; *candidum* 62; *roqueforti* 61
Peppermint and chocolate chips ice cream 38
Pitching curds 55
Pont L'Éveque cheese 82
Pot cheese 82
Potato, cheese 83; salad 32
Pressing curds 58

Rancid butter 46
Raw milk 22

Recipes using, butter 46; cheese 87–9; yoghurt 32
Rennet 48, 53, 55
Rice pudding 24–5
Rich chocolate ice cream 38
Ripening cheese 60
Roman cheese 83
Roquefort cheese 47, 84
Running the whey 55

Sage cheese 84
Salad dressing 32
Salting, butter 44; cheese 58
Scalding the curds and whey 55
Scotch hands 41, 44–5
Selling, butter 44; cheeses 65; cream 39; milk 24; yoghurt 29–30
Skimmed milk cottage cheese 84
Skimming cream 33–4
Smallholder cheese 85
Soda skim cheese 85
Soups 32
Sour cream 36
Soya milk 31
Starter, cheese 49–50, 55; yoghurt 27–8
Sterilized milk 22
Sterilizing equipment 19
Sweet butter 46

Thermometer, dairy 53
Treacle posset 24
Tuscany sheeps' milk cheeses 86
Tutti-frutti ice cream 37

Ultra heat treated (UHT) milk 22
Untreated milk 22

Vanilla ice cream 37
Vanilla pod ice cream 38
Vegetarian, rennet 48; yoghurt 31

Waxing cheese 59
Welsh cheese pudding 88
Welsh rarebit 88
Wensleydale cheese 87
Whey 55

Yoghurt, basic recipe 27; commercial production 29–30; cookery 31–2; making 26–9; problems 28–9; recipes 32; storage 31
Yoghurt liptauer cheese 87